Adsorption
and
Collective
Paramagnetism

PIERCE W. SELWOOD

University of California, Santa Barbara

ACADEMIC PRESS  1962

New York and London

CHEMISTRY

Preface

My purpose in writing this book is to describe a novel method for studying chemisorption. The method is based on the change in the number of unpaired electrons in the adsorbent as chemisorption occurs. The method is applicable to almost all adsorbates, but it is restricted to ferromagnetic adsorbents such as nickel, which may be obtained in the form of very small particles, that is to say, to ferromagnetic adsorbents with a high specific surface. While almost all the data used illustratively have been published elsewhere this is the first complete review of the subject.

The book is addressed primarily to readers interested in heterogeneous catalysis and related areas of surface chemistry, surface physics, and physical metallurgy. For that reason there are included a number of definitions, and an elementary introduction to magnetism. But it is hoped that specialists in magnetism and solid state physics may also find here something of value. For that reason there is included an introduction to adsorption phenomena. If one group finds the detailed magnetic descriptions and calculations to be tedious it is to be hoped that the adsorption work will be found comprehensible, and *vice versa*.

The book was read in manuscript by Dr. Robert P. Eischens and several of his associates, and by Dr. C. A. Neugebauer as well as by his colleagues Dr. Charles P. Bean and Dr. I. S. Jacobs, all of whom made suggestions for correction and amplification. I am indebted also to Dr. Dirk Reinen and to Dr. Y. Yasumori who read the book in proof and who made numerous suggestions for improvement. It is a pleasure to acknowledge the part played by the various graduate students and post-doctorates who contributed to the discovery and development of the methods described. I trust that their respective contributions are adequately recognized at appropriate places in the text.

No list of acknowledgments would be complete without reference to the help received under contract and grant from the Office of Naval Research, the Office of Ordnance Research, the Signal Corps

v

Engineering Laboratories (U. S. Army), the Army Research Office (Durham), the Petroleum Research Fund of the American Chemical Society, the National Science Foundation, and the Advanced Research Projects Agency of the Department of Defense through the Northwestern University Materials Research Center. Without the help of these agencies the work on which this book is based would not have been possible.

September, 1962 P. W. S.

Contents

CHAPTER V

Saturation Data

CHAPTER VI

The Measurement of Magnetization at Low H/T

CHAPTER VII

Magnetization-Volume Isotherms for Hydrogen

CHAPTER VIII

Hydrogen Sulfide, Cyclohexane, Cyclohexene, and Benzene

CHAPTER I

Chemistry of Solid-Vapor Interfaces

1. Introduction

If a surface is exposed to a vapor, molecules of the vapor may adhere to the surface. If this adhesion involves some kind of electronic interaction, the process is referred to as chemisorption. But if no such interaction occurs, the process is called physical, or van der Waals, adsorption.

Until recently there were no very satisfactory methods for studying, or even for recognizing, the occurrence of electronic interaction between adsorbent and adsorbate. There arose, consequently, a tendency to use the term chemisorption for examples of adsorption in which the heat liberated is in excess of about 10 kcal/mole of adsorbate. Physical adsorption was said to occur for cases in which the heat of adsorption is about one order of magnitude less, that is to say, about the same as the latent heat of vaporization of the adsorbate.

There are difficulties in connection with these definitions of chemisorption and physical adsorption. For many systems which obviously involve chemical change, and hence electronic interaction, the heat of adsorption diminishes with increasing surface coverage. Although the reasons for this important phenomenon remain obscure, it is almost certain that chemisorption may occur under circumstances where the heat of adsorption is less than 5 kcal/mole; and evidence has been presented to show that *endothermic* chemisorption may occur. Conversely, there is evidence that in some cases, as of vapors taken up by zeolitic adsorbents, the apparent heat of adsorption (or desorption) is substantially in excess of the latent heat of vaporization, even though other evidence suggests that nothing more than physical adsorption is taking place.

1

Prior to about 1955 the chief emphasis in studies of adsorption processes was on obtaining kinetic and thermodynamic data, by analysis of which a substantial number of theoretical and practical advances became possible. But since that time attention has shifted to new physical methods which give us more direct insight concerning certain aspects of the over-all adsorption process. So far as the study of useful catalysts is concerned, the two physical methods which have thus far produced most information, and which give promise of producing much more, are the method of infrared absorption spectroscopy as applied to adsorbed molecules, and the magnetic method as applied to ferromagnetic adsorbents.

It is the purpose of this review to present in detail the present status of the magnetic method. The primary objective of this method is to provide answers to the following questions: First, what happens to the adsorbed molecule? Second, what kind of surface bond is formed? And third, what happens to the adsorbent? Such an objective needs no apology. These questions and their answers are of major importance in heterogeneous catalysis, in corrosion, in lubrication, in chromatography, and wherever one surface meets another.

2. The Nature of Chemisorption

The magnitude of the heat of adsorption is only one of several criteria by reference to which we may compare physical adsorption with chemisorption. In this section we shall consider other characteristics peculiar to, or at least more pronounced in, the one as opposed to the other.

It will be noted that physical adsorption is a phenomenon which is related to the boiling point of the adsorbate under the pressure prevailing near the surface of the adsorbent. Physical adsorption is negligible except at temperatures near or below the boiling point of the adsorbate; but chemisorption may have a more complicated dependence on temperature. There appear to be well-established cases, as of hydrogen on nickel, in which chemisorption is unimportant at quite low temperatures, reaches a maximum at intermediate temperatures, and diminishes once more at higher temperatures. Certainly it may be said that chemisorption often occurs at temperatures far above those at which appreciable physical adsorption may be detected.

Physical adsorption is often stated (probably erroneously) to be independent of any specific action on the part of the adsorbent. That is to say, in this view the total volume of vapor which may be taken up by any adsorbent depends, in the case of physical adsorption, solely on the temperature and on the total available surface of the adsorbent. But the situation for chemisorption is quite different—specific activity of the adsorbent is the rule rather than the exception. In proof of this statement it is sufficient to point out the strong chemisorption of molecular hydrogen by some metals, of which nickel is one, as contrasted with the negligible adsorption of hydrogen by copper under identical experimental conditions.

It will be noted also that it is quite doubtful if chemisorption may ever proceed beyond the stage of a monolayer of atoms or molecules, as the case may be, on the surface. But for physical adsorption multilayer adsorption is not only possible but will always occur as the pressure of the vapor phase adsorbate becomes appreciable. It will, of course, be clear that physical adsorption may, and frequently does, occur over a chemisorbed monolayer, provided that experimental conditions are appropriate.

It is not infrequently stated that physical adsorption is a rapid process but that chemisorption may be slow. This is a criterion which must be approached with caution. The rate of physical adsorption may, it is true, be limited only by the rate of diffusion to the surface. But if that surface is less accessible by reason of, for instance, being inside a pore of molecular dimensions, then not only will the apparent rate of adsorption be diminished, but Knudsen flow may render the rate almost imperceptible. There is, on the other hand, a mass of evidence tending to show that chemisorption requires, or may require, an activation energy. This implies that the rate of chemisorption should be measurable. Unfortunately, we know so little about the chemisorption process that it is still impossible to say categorically that all chemisorption must involve an activation process. Certainly there are cases in which true chemisorption (as shown by unquestionable electronic interaction) doubtless occurs but for which no rate has been measured. It is the custom to speak of these as "nonactivated" chemisorption, but whether this distinction is valid remains to be seen. It seems to this writer that too frequently the rate of a straightforward "old-fashioned" chemical reduction, as of manganese dioxide by hydrogen, has been needlessly interpreted in terms of an activated adsorption.

Fortunately there are some criteria in which one may have more than a little confidence.* If it may be shown that an adsorbed molecule has suffered some chemical change, such as occurs in the hydrogen-deuterium exchange reaction, or in the self-hydrogenation of ethylene on nickel, then it is difficult to escape the conclusion that some electronic interaction of adsorbent and adsorbate, and hence chemisorption, has taken place. The strictly "chemical" approaches continue to be among the most powerful methods at the disposition of the investigator. To these we may add some physical methods, especially infrared absorption spectroscopy, which make it possible to gain direct information concerning the structure of adsorbed molecular species. There have also been developed methods, of which the magnetic method is one, which make possible direct observations on the electronic state of the adsorbent and on the number and nature of the chemisorption bonds. In true chemisorption there must always be some structural change in the adsorbed molecule and there must always be some change in the electron distribution in the adsorbent. In favorable cases these changes may both be observed, and both are subject to quantitative investigation. It may be said that changes such as these also occur during physical adsorption processes. This is correct, but in physical adsorption the magnitude of the changes is trifling as compared with those in true chemisorption.

3. Experimental Methods in Surface Chemistry[3]

In this section we shall review several experimental methods which are available for elucidating the mechanism of adsorption. The plan will be to describe the method briefly and to state the kinds of information which it provides. Our purpose in this is to provide some background for comparison of the information obtainable by the magnetic method with that from other methods. If some searching techniques have been slighted, the reason is that they do not yield results with which we may directly compare the

---

* Although perhaps a little less than formerly. De Boer[1] refers to a remarkable example in which mercury isotopes in the two aromatic complexes Ar'—Hg—Ar' and Ar—Hg*—Ar were found to exchange without the formation of mixed complexes.[2]

conclusions based on the changes of magnetization to be described in later chapters. The references given in connection with this section will, in almost every case, be restricted to review articles.

It may seem odd that we start by referring to a purely theoretical approach to the problem. But no account of adsorption can now be complete without at least a passing reference to wave mechanics[4] and electron theory[5] as applied to the surface bond. The problems involved, particularly in connection with the former, are indeed formidable. But these methods may ultimately yield the answers which experimental methods have not been able to provide.

The first of the truly experimental methods to be mentioned are those involving adsorption equilibria and the rates of their attainment.[6,7] These are among the oldest and most fruitful of pursuits in surface chemistry. The experiments consist of volumetric (or occasionally gravimetric) measurement of the quantity of vapor sorbed, at several pressures and temperatures. Such equilibrium studies make it possible to find the surface area of the sorbent and the heat of sorption, or to gain information concerning such molecular processes as may occur on the surface. In recent years studies on the energy of adsorption and lateral interaction have become increasingly sophisticated.

Physical adsorption is not without effect on the physical properties of the adsorbent. This is true in, for instance, the perceptible dimensional changes in porous glass during the adsorption of inert gases,[8] and in the small but measurable changes in the contact potential of metal films caused by adsorbed argon.[9] It is also true that the adsorbent may affect the structure of the adsorbed layer or layers. But, in general, the information derived from thermodynamic and kinetic studies of physical adsorption processes will prove to be of minor importance for our main purpose.

When we go to chemisorption the situation is just the opposite.[10] A wealth of experimental data shows that the heat of adsorption is related to the strength of the adsorbent-adsorbate bond, that it generally diminishes with increasing surface coverage, and that these phenomena are both of prime importance in heterogeneous catalysis. Many attempts have been made to relate the activation energy of adsorption to the kind of chemical change required to complete the adsorption process. Many examples of chemisorption are dissociative in nature, and the heat of dissociation is often indicative of the kinds of dissociation which are thermodynamically possible; the form of the adsorption isotherm is different for different modes of adsorption. Dissociative adsorption (as of molecular hydrogen) into two molecular species, both of which are adsorbed, involves reaction between a vapor molecule and two (or in some cases more) surface sites, while dissociative adsorption (as of nitrous oxide) may yield one species which is adsorbed and one which is not. Similarly, while one species may react with another on the surface, lowering the temperature may halt such reaction and make it possible to estimate the surface coverage produced by dissociation, or even fragmentation, of an adsorbed molecule. From considerations such as these we may gain much information, and we shall make frequent reference to such methods later.

The related, and potentially useful, method of accommodation coefficients[11] has not yielded results of major importance, but this is perhaps due in part to the difficulty of precise measurement.

A careful study of reaction products,[12,13] including the use of isotopes, is still a method of prime importance to surface chemistry, and this is true in spite of the development of powerful physical methods. In fact it is a tribute to the older methods that few surprises have emerged from the application of sophisticated modern techniques. The fact that hydrogen and deuterium may suffer exchange on a catalyst surface is certainly strong (if not quite conclusive) evidence establishing rupture of the hydrogen-hydrogen bond through the agency of the surface. Equally significant is the formation of ammonia from molecular hydrogen plus molecular nitrogen. To such relatively simple examples we may add many others, some of which, involving complicated arguments relative to the stereochemistry of the adsorbate, reveal possible modes of adsorption and interaction through study of the reactions of exchange, isomerization, and the like.[14,15] One of the most useful of such techniques (for our present purposes) is the refinement by Kemball[16] of deuterium exchange for determining the number of dissociated hydrogens per molecule of adsorbate. Cyclohexene is, for instance, chemisorbed in the room temperature region on nickel by dissociative adsorption leading to the formation of two nickel-hydrogen bonds per molecule of cyclohexene. The information thus obtained lends itself readily to direct comparison with the magnetic method under conditions which appear to be identical. The agreement, as we shall in due course see, is gratifying.

We turn now to the physical methods, some of which give information concerning the adsorbed molecule, and others of which give information concerning such changes as may have occurred in the adsorbent. Some methods give a measure of information relative to both adsorbent and adsorbate simultaneously. We shall group together electron microscopy and electron diffraction. In recent years electron microscopy has reached surprising limits of resolution, but thus far the limit still lacks about one order of magnitude for our purpose. *Not any more*

The method of electron diffraction extends the limit of resolution.[17,18] It is true that we are restricted to rather specialized conditions. These include quite high vacuum* and extraordinary care to obtain clean surfaces. But it is possible to use single crystals with relatively large, defined faces. Under these conditions, and with low-speed electron beams, one may "observe" adsorbed molecules and make useful interpretations concerning their mode of attachment and their distribution on the surface. In favorable cases it is possible to draw some conclusions concerning the mutual reactions of adsorbed molecules. We shall not have many occasions on which to compare low-speed electron diffraction data with the magnetic method, but we shall have some.

The method of infrared absorption spectroscopy[19] is, for our purposes, in quite a different category (see Fig. 1). We shall examine the method in some little detail. At first glance it would appear to be impracticable to observe the absorption spectrum of a monolayer of adsorbed molecules; but under favorable conditions this may be done. The restriction is that the adsorbent must be present in very small particles, that is to say, with a high specific surface. This is a definite advantage because it places the method in the class of using the same, or similar, samples as those to which we shall have reference in the magnetic method. The underlying reason for this is obvious. We are attempting to measure the properties of an adsorbate which is inevitably present in rather small proportion relative to the adsorbent. In order to obtain a useful signal we must make the proportion of adsorbate as large

* We shall use the term "high vacuum" to mean pressures substantially lower than 10^{-6} mm Hg.

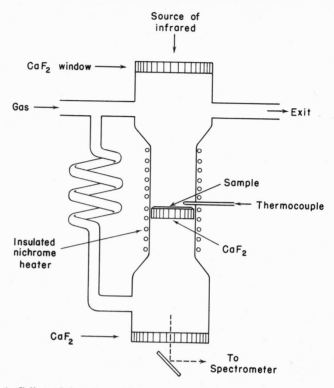

FIG. 1. Cell used for observation of the infrared absorption spectrum produced by molecules adsorbed on metals and oxides of interest in heterogeneous catalysis (after Eischens).

as possible and this may only be done by enlarging the specific surface of the adsorbent. Inasmuch as this is also the condition required, more often than not, for efficient catalysis, we find the method of absorption spectroscopy, together with the magnetic method, to be almost unique among the physical methods in their ability to deal with practical catalysts.

The infrared absorption spectrum of adsorbed molecules is obtained as rather poorly defined, but nevertheless recognizable bands (see Fig. 2). These are observed by passing the beam through a thin deposit of supported catalyst, as of nickel on silica gel; the total mass of metal adsorbent being only a fraction of a milligram. Various gaseous adsorbates may be admitted to the sample. The relatively large dead-space must be kept at low pressure, and it is obviously difficult to make a very accurate estimate of the

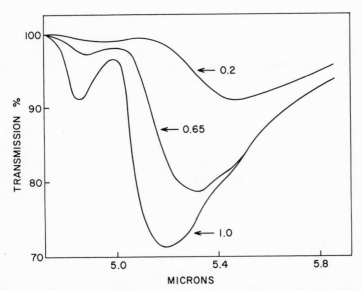

FIG. 2. Infrared absorption bands of carbon monoxide adsorbed on platinum at several levels of surface coverage, as indicated (after Eischens and Pliskin).

amount of vapor actually taken up.

The absorption bands are identified, for the most part, by comparison with the known spectrum of gaseous molecules. It is, for instance, known that the carbon-oxygen stretching frequency in carbonyls produces a band at 4.9 microns for "linear" $M=C=O$ systems, but at 5.3 microns for "bridged" structures, as shown below. Both of these are found in carbonyls such as $Fe_2(CO)_9$.

The observation of bands at these frequencies is taken as proof of the mode of attachment of the chemisorbed molecule.

The method has yielded a wealth of interesting information. We shall see later that the method yields results which are sometimes not in complete agreement with those from other methods. A criticism sometimes heard concerning the absorption spectrum method is that the bands associated with a group in a vapor phase molecule should not necessarily be the same as the bands in a group attached to a metal atom which is itself part of a metal surface. To give a specific example—is it correct to say that the bands associated with a C—H group in ethylene are necessarily the same as those for the same group when the carbon is not only attached to a nickel atom, but when that nickel atom is itself part of the complicated system found at a metal surface?

These remarks should not be construed as a general criticism of what has proved to be a powerful tool in surface chemistry.[20] Rather they are meant to show one possible direction in which we may look in an effort to reconcile a few discrepancies. Concerning these we shall have more to say later.

The measurement of magnetic susceptibility of an adsorbed gas is a method which, in the hands of a few devoted followers, has yielded useful results.[21] The method consists of measuring the susceptibility of an adsorbent before and after admission of a gas which may be taken up by the surface of the adsorbent. The method is, in general, restricted to paramagnetic gases such as oxygen or nitric oxide because these are the only adsorbates which might be expected to show measurable changes of susceptibility after adsorption. In this way it is possible, for instance, to follow the conversion of paramagnetic molecular oxygen to diamagnetic oxygen while it is adsorbed on graphite.

It is doubtful if this method could be refined to a degree such that it would be sensitive enough to detect changes in diamagnetic molecules. Actually there is no particular reason for attempting such an improvement in sensitivity because the same purpose is achieved, with vastly improved precision, by the study of nuclear magnetic resonance in adsorbent-adsorbate systems. The information so gained is related chiefly to the electronic environment of protons in, or derived from, the adsorbate.

These magnetic susceptibility methods should not be confused with those in which the magnetic changes occur in the adsorbent. These latter form the principal subject of this review.

We turn now to those methods in which attention is directed primarily to the adsorbent rather than to the adsorbate. The field-emission microscope actually belongs to this class because it uses changes of work function; and the first method to be described below, that of surface dipole measurement, certainly belongs to both classes.

Adsorption of a molecule on a surface creates a surface dipole which may be measured with fair precision, and the sign and magnitude of which give some information concerning the mechanism of adsorption. The surface dipole is related to the work function and, in the classical work of Langmuir, was measured simply as the change of contact potential between two wires, one of which serves as a thermionic emitter.

Methods currently in use for the study of the several electron-emission and surface potential phenomena include modifications of the original Langmuir method, the field-emission microscope, photoelectric methods, and a variety of methods which, directly or indirectly, measure the contact potential difference between a surface under investigation and a reference surface.[22] For our purposes the most useful methods are the photoelectric work function method (see Fig. 3) and the capacitor method. In the former

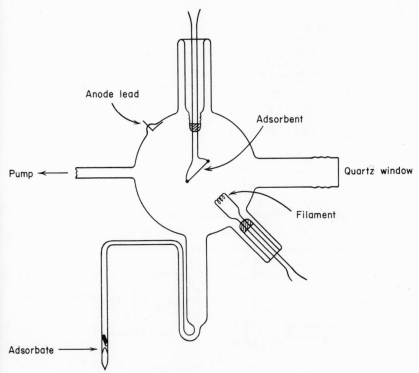

FIG. 3. Apparatus for measuring the change of photoelectric work function caused by an adsorbed vapor. The filament is used to produce the adsorbent film (after Suhrmann).

a beam of monochromatic light strikes a metal cathode which may be a foil or evaporated film. The photocurrent produced per unit light intensity may be measured for different frequencies of light and this current is changed if molecules are adsorbed on the cathode. In some cases adsorbed molecules increase the work function and in other cases the reverse is true. The results are also strongly dependent on surface contamination. For ionic adsorption the change of work function may possibly be related to the establishment of surface dipoles. But for covalent adsorption it is doubtful if any real understanding may yet be gained. There have, however, been a sufficient number of studies made to show that the work function is potentially a valuable aid in surface chemistry.

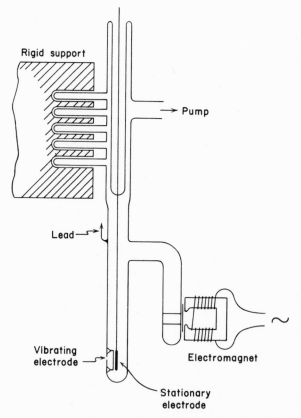

FIG. 4. Vibrating condenser method for measuring surface dipoles (after Mignolet).

The measurement of surface dipoles by the vibrating condenser plate method has been much studied (see Fig. 4). Here again the sample must be a foil or film such that the surface is well defined. The reference plate must be inert to the vapor molecules introduced. It is possible to measure the sign and magnitude of the dipole produced and the method appears, at first glance, to be one of considerable power. But interpretation of the data is far from simple. Even for such a case as hydrogen on platinum one may often hear it said that chemisorption would be easier to understand if surface dipoles had never been measured. This is, of course, a gross exaggeration, but we shall not find it easy to compare the surface dipole method with others to which we shall have reference.

In recent years field emission microscopy has reached astonishing limits of resolution (see Fig. 5). In this method, as is well

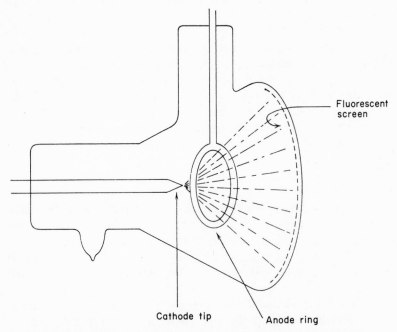

FIG. 5. Principle of the field-emission microscope (after Müller).

known, a wire is etched to a very sharp tip of radius 10^{-5} to 10^{-4} cm. The wire is placed in a coated glass envelope which serves as anode and fluorescent screen. On application of a potential of 10^4

volts or higher between tip and screen, emission of electrons occurs if the tip is negative. The work functions of different crystal faces exposed on the tip vary slightly from face to face and this produces an image of the tip on the screen. Differences in work function also result from the presence of adsorbed molecules on the surface of the tip. If the tip is made the anode and hydrogen, at about 10^{-3} mm pressure, is placed in the device it is found that hydrogen ions leave the tip, with a consequent improvement of resolution in the pattern observed.

The ability to photograph arrays of individual atoms and to observe directly the motions of adsorbed molecules on those atoms is certainly one of the major achievements of science.[23] It might, therefore, be thought that this would be the ultimate method with which we might compare all others. It is true that the field emission microscope has yielded valuable insight into the kinetics of adsorption, and especially with respect to surface mobility.[24] It has also proved useful in establishing the preferential adsorption of molecules on certain crystal faces. Nevertheless, the special conditions required in the microscope, the critical dimensions of the tip, and some uncertainties in interpretation seem to be the reasons why this method has not as yet made any very important contributions to the practice of catalysis. The same is, of course, true of less spectacular methods, of which the magnetic method is one.

Electrical conductivity measurements, as an aid to understanding adsorption processes, are made, for the most part on evaporated films prepared under high vacuum.[25] Such films show appreciable changes in conductivity as adsorption occurs as, for instance, in the case of hydrogen on nickel. The conductivity is often found to increase although, for certain conditions of admission and surface coverage, a decrease of conductivity occurs. These results are interpreted as showing electron transfer from adsorbate to adsorbent for increasing conductivity, and vice versa for decreasing conductivity.

With the experimental observations we can have little quarrel except to point out that the magnitude of the effect and even the sign depend to some degree on the temperature at which the film has been outgassed. But interpretation of the data poses a difficult problem. The structure and properties of such films are themselves subject to much debate. How to relate conductivity changes to mechanism of adsorption is a problem which has not yet been resolved. It is often stated that an increase of conductivity results from electron transfer from adsorbate to adsorbent, but concerning this there is certainly no agreement and no satisfactory theoretical basis. A growing impression seems to be that on properly cleaned metal surfaces all adsorbate molecules increase the work function and decrease the conductivity; but concerning the possible truth of this statement we do not feel competent to express an opinion.

This section will be concluded by reference to the method of X-ray K absorption edge spectrometry.[26] For a number of atoms, some of which are of prime interest in catalysis, the shape of the K absorption edge may be determined with considerable precision. This edge is determined in part by the chemical environment of the atom under investigation. This is to say that manganese as a metal gives an edge pattern (see Fig. 6) recognizably different from

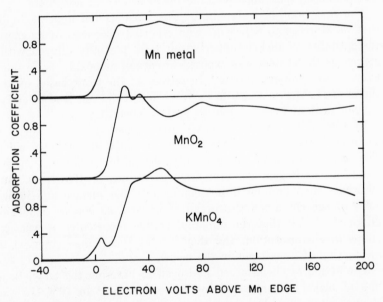

FIG. 6. X-ray K absorption edges produced by manganese in three different atomic environments (after Van Nordstrand).

that of manganese in manganese dioxide or in potassium permanganate. In brief, the method helps us to identify the oxidation state and the coordination of a metal atom.

The experimental determination of adsorption edges is somewhat tedious, and the theory of the changes which occur is still in a somewhat rudimentary state. But the method may be used on samples which have direct application in heterogeneous catalysis, and the results are therefore directly comparable with those obtained by the infrared absorption method and by the magnetic method. The method, together with a possible related use of high-resolution X-ray fluorescence spectroscopy, has considerable in its favor.

4. Introduction to the Magnetic Method

It may assist the reader if this section is devoted to a brief outline of the magnetic method for studying chemisorption as it has developed in the author's laboratory. The method is based on the change of magnetization which may be observed in certain adsorbents as their surfaces take up molecules of adsorbate.

The first published account of this phenomenon appears to be that of Morris and Selwood[27] who reported decreases of magnetic susceptibility of nickel supported on (and partially alloyed with) copper as the system was exposed to carbon monoxide, or to mercury, or to hydrogen sulfide.* In retrospect the work based on the thesis of H. Morris seems to be valid, but it did not lead to active development of a useful tool in surface chemistry.

In 1948 Dilke et al.[28] reported that dimethyl sulfide causes a decrease in the magnetic susceptibility of palladium metal powder. The loss of magnetic susceptibility shown by palladium on exposure to hydrogen has been known for nearly 100 years. But hydrogen on palladium is not confined to the surface and it has more of the characteristics of an alloy such as copper in nickel. Dilke et al. felt that the dimethyl sulfide represented a true example of chemisorption, and they related the change observed to the pairing of d electrons in the metal.

The action of chemisorbed hydrogen in lowering the magnetization of nickel was reported by Selwood et al.[29] in 1954. It was at once clear that this observation offered an opportunity for development of a unique method in surface chemistry. This development proceeded fairly rapidly and may now be considered virtually complete. It was at first thought that the ferromagnetism of the

---

* About 1949 the late Dr. Otto Beeck told the author that he had observed a change of magnetization in nickel under similar circumstances. To the best of the author's knowledge this observation was never published.

nickel would make such effects impossible to observe or, if observed, impossible to interpret. Actually, the ferromagnetism greatly facilitates the observations and does not make the interpretation too difficult.

If an adsorbent possesses an unfilled d band or, to put it another way, possesses unpaired d electrons in its orbitals, then the electronic interaction leading to formation of a chemisorption bond might be expected to alter the filling of the d band. The saturation magnetic moment depends on the number of unpaired electrons. Hence, we may expect to observe a change of saturation magnetization as chemisorption proceeds.

The measurement of saturation magnetization in a ferromagnetic substance is not necessarily easy. Whether the magnetization change observed at moderate fields and temperatures bears any simple relation to the change of saturation magnetization is something which could scarcely have been predicted prior to development of the Stoner-Néel theory of magnetization in small particles. But Néel's contribution to the theory was developed in 1949; and it is obvious that small particles are required if the adsorbent is to have a ratio of surface to volume high enough so that an appreciable fraction of the adsorbent atoms are in a position to be affected by adsorbate molecules.

All these conditions are met in typical reduced nickel-silica hydrogenation catalysts. The method is also applicable to cobalt and, presumably, to iron. There are few restrictions on the kinds of adsorbate for which the method may be used.

REFERENCES

1. J. H. de Boer, "The Mechanism of Heterogeneous Catalysis," p. 11. Elsevier, Amsterdam, 1960.
2. O. A. Reutow, Lecture at the 17th International Union of Pure and Applied Chemistry, Munich, 1959.
3. P. M. Gundry and F. C. Tompkins, *Quart. Revs.* **14**, 257 (1960).
4. T. B. Grimley, *Advances in Catalysis* **12**, 1–30 (1960).
5. T. Wolkenstein, *Advances in Catalysis* **12**, 189–264 (1960).
6. H. E. Ries, Jr., *in* "Catalysis," Vol. I, Fundamental Principles (Part 1) (P. H. Emmett, ed.), pp. 1–30. Reinhold, New York, 1954.
7. K. J. Laidler, *in* "Catalysis," Vol. I, Fundamental Principles (Part 1) (P. H. Emmett, ed.), pp. 75–118. Reinhold, New York, 1954.
8. D. J. C. Yates, *Advances in Catalysis* **12**, 265–312 (1960).
9. J. C. P. Mignolet, *Discussions Faraday Soc.* **8**, 105, 326 (1950).
10. B. M. W. Trapnell, "Chemisorption." Academic Press, New York, 1955.

11. D. D. Eley, *Advances in Catalysis* **1**, 160 (1948).
12. C. Kemball, *Bull. soc. chim. Belges.* **67**, 373–398 (1958).
13. C. Kemball, *Advances in Catalysis* **11**, 223–262 (1959).
14. A. A. Balandin, *Advances in Catalysis* **10**, 96 (1958).
15. R. L. Burwell, Jr., "Techniques nouvelles en Catalyse hétérogene," pp. 25–46. Soc. des Editions Tech., Paris, 1960.
16. A. K. Galwey and C. Kemball, *Actes 2e congr. intern. catalyse, Paris, 1960* **2**, 1063 (1961).
17. R. E. Schlier and H. E. Farnsworth, *Advances in Catalysis* **9**, 434 (1957).
18. L. H. Germer, *Advances in Catalysis* **13**, 191 (1962).
19. R. P. Eischens and W. A. Pliskin, *Advances in Catalysis* **10**, 2 (1958).
20. V. Crawford, *Quart. Revs.* **14**, 378 (1960).
21. P. W. Selwood, "Magnetochemistry," p. 279. Interscience, New York, 1956.
22. R. V. Culver and F. C. Tompkins, *Advances in Catalysis* **11**, 67–131 (1959).
23. E. Müller, *Ergeb. exak. Natur.* **27**, 290 (1953).
24. R. Gomer, *Advances in Catalysis* **7**, 93–134 (1955).
25. R. Suhrmann, *Advances in Catalysis* **7**, 303–352 (1955).
26. R. A. Van Nordstrand, *Advances in Catalysis* **12**, 149–187 (1960).
27. H. Morris and P. W. Selwood, *J. Am. Chem. Soc.* **65**, 2245 (1943).
28. M. H. Dilke, D. D. Eley, and E. B. Maxted, *Nature* **161**, 804 (1948).
29. P. W. Selwood, S. Adler, and T. R. Phillips, *J. Am. Chem. Soc.* **76**, 2281 (1954); **77**, 1462 (1955).

CHAPTER II

Introduction to Magnetism

1. Magnetic Phenomena[1]

The reader may find it to be a convenience if there is given a general description of effects related to magnetic fields, so far as these are of concern to our chief purpose. This description will be followed by some definitions, and by some relations between magnetism and atomic structure.

Experience tells us that in the neighborhood of a magnet, or of an electric current as in a piece of wire, space has some unique properties. The properties include the ability to attract a piece of iron and to orient a compass needle. These effects are said to be caused by a magnetic field. A magnetic field has both strength and direction.

Some substances have this property that when placed in a magnetic field which has a gradient of strength from point to point they are repelled to a region of weaker field strength. Such substances, of which water is an example, are said to be diamagnetic. Other substances have the property of being attracted to a region of higher magnetic field strength. Such substances, of which molecular oxygen is an example, are said to be paramagnetic. Some substances are not only strongly attracted to a region of higher field strength, but they themselves have the ability to create quite large magnetic fields, even in the absence of an external field. These substances, of which iron is an example, are said to be ferromagnetic. A few specialized kinds of magnetic behavior will be referred to later, but our chief interest will be with ferromagnetism and paramagnetism, and the border region between them.

It will be of most service to readers concerned with the phenomena of adsorption if we use the cgs and practical system of

19

units. In this system the unit of magnetic field strength, H, is given in oersteds. The oersted has the dimensions $cm^{-1/2}g^{1/2}sec^{-1}$.

A body of matter placed in a magnetic field is said to be magnetized. The total magnetic induction in such a body is B, the unit of which is generally given as the gauss, although B and H appear to have the same dimensions. Then the intensity of magnetization or simply the magnetization M, is given by the relation*:

$$B = H + 4\pi M \qquad (2.1)$$

The magnetization is often stated to be in "emu," although the gauss is a perfectly acceptable unit. If H and M differ in direction the addition will be vectorial, but this will not concern us. It will be noted that the magnetization is also the magnetic moment per unit volume. This is true because the moment, as in a rod, is equal to the pole strength times the distance between the poles. The magnetization is equal to the pole strength divided by the cross section of the rod.

The magnetization divided by the density, d, is called the specific magnetization. For this the symbol σ will be used. The magnetic susceptibility (per unit volume) is $\kappa = M/H$ and the susceptibility per gram is $\chi = \kappa/d$. This has units of reciprocal density.

The magnetic permeability (which we shall use rarely, if at all) is the fraction B/H.

2. Atomic Basis of Diamagnetism and Paramagnetism[2]

The magnetic susceptibility of a diamagnetic substance is negative. A typical value is that for water, for which $\chi = -0.720 \times 10^{-6}$ cm^3 g^{-1}. Diamagnetic susceptibilities are independent of field strength. In general they are independent of temperature although there are some exceptions, of which that of graphite is one.

Diamagnetism results from the Larmor precession induced in electronic motion by the applied field. It depends primarily on the average radial charge distribution in the atom and, like an induced electric dipole moment, is present only when the external field is applied.

---

* Origin of the term 4π is based on the convention that unit field strength exists at a point where the force on unit pole is 1 dyne. At a distance of 1 cm from a unit pole the field created by the pole is distributed on the surface of a sphere. The surface area of the sphere is $4\pi cm^2$. Hence 4π lines of force must leave unit pole to create unit field at a distance of 1 cm.

The susceptibility of a paramagnetic substance is positive, a typical value is that for $CuSO_4 \cdot 5H_2O$ for which $\chi = 5.85 \times 10^{-6}$ at 20°C. Paramagnetic susceptibilities are independent of field strength except at very high fields and low temperatures when magnetic saturation is said to occur. For many paramagnetic substances the susceptibility varies inversely as the absolute temperature, T, according to the Curie law:

$$\chi = C/T \qquad (2.2)$$

where C is the Curie constant. More frequently the paramagnetic susceptibility may be represented by the Curie-Weiss law:

$$\chi = C/(T + \Delta) \qquad (2.3)$$

over a considerable range of temperature. The quantity Δ is called the Weiss constant; its origin will be discussed below.*

Paramagnetism arises from unpaired electrons. The atomic magnetic moment so produced is independent of the applied field in the same sense that a permanent electric dipole moment is independent of the applied electric field. The function of the applied field is thus merely one of bringing about a preferred orientation of the atomic magnetic moments already present. Some atomic nuclei have a permanent magnetic moment. This, though important for some purposes, is small compared with the moment associated with unpaired electrons. We shall not be concerned with nuclear moments.

Paramagnetism is, in general, found in so-called "odd" molecules, i.e., those possessing an odd number of electrons. It is also found in a few compounds such as molecular oxygen and the Chichibabin biradical hydrocarbons which, although possessing an even number of electrons, have two of these electrons unpaired. Similarly, paramagnetism is found in atoms and ions of the several transition series of elements, provided that these contain a partially filled d or f electron energy level. The conduction electrons in metals also contribute a modest paramagnetism to metals. This may or may not be overshadowed by the underlying diamagnetism which is, of course, observed in every substance in the presence of an applied magnetic field.

Let the magnetic moment of a particle be μ, and let there be N

* Many authors use the Curie-Weiss law in the form $\chi = C/(T - \Delta)$, there being no agreement as to the sign of the Weiss constant.

such particles present in a sample of unit volume. The presence of an applied field, H, tends to orient these moments and, for very high fields, complete orientation would be approached. The saturation magnetization so produced would be $M_s = N\mu$, but in practice this is difficult to attain because thermal agitation prevents simultaneous orientation of the moments. Under these circumstances the observed magnetization is less than $N\mu$, as given by the Langevin equation:

$$\frac{M}{M_s} = \coth\frac{NH}{kT} - \frac{kT}{H} \tag{2.4}$$

where k is the Boltzmann constant. If the fraction M/M_s is considerably less than unity, we may simplify Eq. (2.4) to read:

$$M = N\mu^2 H/3kT \tag{2.5}$$

which may be rewritten in terms of susceptibility as

$$\kappa = N\mu^2/3kT \tag{2.6}$$

This is, of course, equivalent to the Curie law. It shows the variation with temperature actually found for many paramagnetic substances in readily attainable fields and at temperatures not too close to the absolute zero.

Quantum theory yields the same expression for paramagnetic susceptibility. It is convenient to express atomic magnetic moments in units of the Bohr magneton, β.

$$\beta = eh/4\pi mc = 9.27 \times 10^{-21} \text{ erg gauss}^{-1} \tag{2.7}$$

where e is the electronic charge, h is Planck's constant, m is the mass of an electron, and c the velocity of light. Equation (2.6) then becomes

$$\chi_A = A\mu_A^2\beta^2/3kT \tag{2.8}$$

where χ_A is the susceptibility per gram-atom, A is Avogadro's number, and μ_A is the magnetic moment per atom expressed in Bohr magnetons.

From Eq. (2.8) we may find the number of Bohr magnetons per atom or ion from susceptibility measurements conducted, preferably, over a range of temperature, and with appropriate corrections for the underlying diamagnetism. The moment is given by:

$$\mu_A = 2.84[\chi_A(T + \Delta)]^{1/2} \tag{2.9}$$

Some typical values for μ_A are: Cu^{2+}, 1.7; O_2, 2.8; Cr^{3+}, 3.8; Fe^{3+}, 5.9; and Gd^{3+}, 7.9.

The number of Bohr magnetons depends not only on the number of unpaired electrons in the atom or ion but also on the way in which the contributions of electron spin and orbital angular momentum may be combined. If only the spin component need be considered, then:

$$\mu_A = 2[S(S + 1)]^{1/2} \tag{2.10}$$

where S is the spin quantum number. This applies to a fair number of commonly encountered substances and is known as the "spin-only" formula. (The number of unpaired electron spins is equal to $2S$.) For typical organic free radicals such as triphenylmethyl, with one unpaired electron, the moment is indeed about 1.7 Bohr magnetons. For Mn^{2+} $\mu_A = 5.9$, corresponding to five electron spins. In Table I there are given the number of d electrons,

TABLE I

MAGNETIC MOMENTS OF IONS OF THE FIRST TRANSITION SERIES

Ion	$3d$ electrons	Unpaired electrons	$2[S(S + 1)]^{1/2}$	μ_A (obs)
Sc^{3+}, Ti^{4+}, V^{5+}	0	0	0.00	0.0
Ti^{3+}, V^{4+}	1	1	1.73	1.8
V^{3+}	2	2	2.83	2.8–2.9
V^{2+}, Cr^{3+}, Mn^{4+}	3	3	3.87	3.7–4.0
Cr^{2+}, Mn^{3+}	4	4	4.90	4.8–5.1
Mn^{2+}, Fe^{3+}	5	5	5.92	5.2–6.0
Fe^{2+}	6	4	4.90	5.0–5.5
Co^{2+}	7	3	3.87	4.4–5.2
Ni^{2+}	8	2	2.83	2.9–3.4
Cu^{2+}	9	1	1.73	1.8–2.2
Cu^+, Zn^{2+}	10	0	0.00	0.0

the spin-only moment, and the observed moment for several ions of the first transition series.

But for many substances the observed moment depends on both spin and orbital contributions, in which case:

$$\mu_A = g[J(J + 1)]^{1/2} \tag{2.11}$$

where g is the Landé splitting factor given by

$$g = \frac{1 + J(J + 1) + S(S + 1) - L(L + 1)}{2J(J + 1)} \tag{2.12}$$

J and L being the total and orbital quantum numbers, respectively. Equation (2.11) is most useful for the rare earths and the actinides where, for the most part, the f electrons are shielded from external electric fields and other effects. But for the common elements of the first transition series the spin-only formula is quite useful for nearly all ions except Co^{2+}. In these ions the orbital contribution to the magnetic moment is said to be "quenched." Actually, the influence of neighboring atoms and ions may be quite complicated and, as it will concern us later, the matter will be considered in some little detail.

The magnetic moment of an atom isolated from other atoms may be predicted from Eq. (2.11), but few atoms occur in such an environment. In a solid or liquid those atoms or ions possessing an incomplete inner energy level, or participating in an incomplete d band, are subject to several influences: first, the presence of adjacent ions or oriented polar molecules has two important effects. One is that the L and S coupling may be altered in such a way that the usual J values no longer properly define the several states, and the L sublevels may be greatly altered. The other is that in an electric field the orbital contribution to the magnetic moment may be completely, or almost completely, eliminated. This latter effect is the reason that a considerable number of ions with incomplete d shells have magnetic moments predictable by the spin-only formula. A few elements, of which Co^{2+} is a notable example, have moments which vary considerably depending upon the symmetry and charge intensity of the coordination sphere, or ligands, surrounding the ion. This is the area of application of the crystal field theory. In a considerable number of such cases the susceptibility may be represented over a fairly wide temperature range by the Curie-Weiss law but it is doubtful if, in such cases, the Weiss constant has any real significance.

3. Antiferromagnetism

There is another important effect caused by the influence of paramagnetic ions upon each other. A substance in which the density of unpaired electrons is relatively high is said to be "magnetically concentrated." We may start with a dilute solid solution of, say, Cr_2O_3 in the isomorphous, diamagnetic solvent Al_2O_3. At low Cr^{3+} concentrations the paramagnetic ions are relatively far apart and exert minimum influence on each other. As the concentration is raised we find that the moment of each Cr^{3+} ion begins to have an effect on all the other Cr^{3+} ions in the immediate neighborhood. This has the effect of increasing the Weiss constant, which is practically zero for the dilute solid solution, but which may become quite large as pure Cr_2O_3 is approached. This effect is due to exchange interaction between adjacent paramagnetic ions. Exchange interaction may be such that electron spins are ordered in the same direction—a condition which leads to ferromagnetism, which will be discussed in the following section; or it may be such that the spins are ordered in opposite directions. This latter case is our present concern. It is called antiferromagnetism. Most oxides of the transition elements, and many other compounds, are antiferromagnetic. The phenomenon is more common than is ferromagnetism. The rare earth oxides, for obvious reasons, show only traces of antiferromagnetism, although negative exchange interaction is quite strong in uranium dioxide.

Antiferromagnetics are characterized by a susceptibility which follows the Curie-Weiss law at elevated temperatures but which abruptly turns lower below the Curie temperature. (This is sometimes called the Néel temperature). The Curie temperature for MnO is sharply defined at $122°K$, while that for Cr_2O_3 is poorly defined at about $310°K$. It is of considerable interest to workers in the field of heterogeneous catalysis that supported oxides such as Cr_2O_3 on gamma-Al_2O_8, and also the familiar high specific surface gels such as chromia aerogel, tend to be normal paramagnetics rather than antiferromagnetics. The reason for this is doubtless the degree of attenuation which places each paramagnetic ion in an environment with much less than its normal coordination of paramagnetic neighbors.[3]

It remains in this section to say just a word or two concerning the paramagnetism of conduction electrons in metals. This is quite small. It is estimated in sodium metal, for instance, to be about 0.6×10^{-6} per gram. After several corrections this yields a calculated susceptibility of almost 0.5×10^{-6} as compared with the observed value of 0.7×10^{-6}. Conduction electron paramagnetism should not be confused with the larger paramagnetism shown by some metals which have an incomplete d shell; of these, palladium with a susceptibility of about 5.25×10^{-6} is a notable example. Table II gives a few susceptibilities to which we may have refer-

TABLE II
MAGNETIC SUSCEPTIBILITIES AT ROOM TEMPERATURE

Substance	$\chi \times 10^6$	Substance	$\chi \times 10^6$
Alumina, Al_2O_3	−0.3	Magnesium, Mg	0.25?
Aluminum, Al	+0.6	Mercury, Hg	−0.17
Argon, Ar	−0.5	Nitrogen, N_2	−0.43
Calcium, Ca	+0.7	Oxide, ion, O^{2-}	−0.75
Copper, Cu	−0.83	Oxygen, O_2	+107.8
Graphite, C	−7.8[a]	Platinum, Pt	+1.0
Gold, Au	−0.15	Silica, SiO_2	−0.5
Hydrogen, H_2	−2.0	Silver, Ag	−0.2
Hydroxide ion, OH^-	−0.70	Zinc, Zn	−0.17

[a] Crystalline graphite shows a remarkable example of magnetic anisotropy. The susceptibility of the powder, which is actually an average of the three principal susceptibilities as measured along the magnetic axes, is strongly dependent on particle size. The value given is for coarsely powdered crystals.

ence later. Extensive tables of susceptibilities will be found in "Constantes Sélectionnées."[4]

4. Ferromagnetism[5]

Some substances including iron, cobalt, nickel, and a few oxides and other compounds have a unique kind of magnetic behavior. If placed in a magnetic field they become magnetized to an extraordinarily large degree for quite moderate intensities of the applied field. But as the external field is raised, the magnetization approaches a limit as shown in Fig. 7. Substances behaving in this fashion are said to be ferromagnetic.

Ferromagnetic substances also show unique behavior when the magnetization is measured as a function of temperature. It will be found that as the temperature is raised the magnetization falls slowly and then, at a definite temperature, falls rapidly almost, but not quite, to zero. The temperature at which abrupt loss of magnetization occurs is called the Curie temperature, T_c. For nickel this is 631°K. Some little distance above the Curie temperature the substance may behave as a typical paramagnetic following the Curie-Weiss law with a Weiss constant not too different numerically from the Curie constant.

The reason that ferromagnetic substances act as they do is that below the Curie temperature they are magnetized even in the

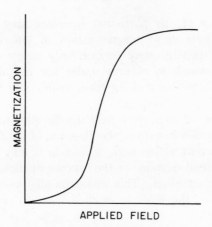

FIG. 7. Magnetization as a function of field for a typical ferromagnet.

absence of an external field. It was first suggested by Pierre Weiss that in ferromagnetics there is a very large internal field (of the order of 10^6 gauss), that this field is proportional to the magnetization, and that it causes the magnetization to approach the maximum possible for complete parallel orientation of the atomic moments. In paramagnetics the orientation of magnetic dipoles is opposed by thermal agitation. But in ferromagnetics the electron moments are locked together in the same direction so that they act cooperatively. The whole group acts as one very large magnetic moment. In such a group the electron spins are held in parallel alignment by quantum-mechanical exchange forces. The situation is then not unlike that already mentioned for antiferromagnetic substances except that in antiferromagnetics the spins are locked together in antiparallel alignment whereas in ferromagnetics the alignment is parallel. This parallel alignment persists against thermal agitation until it finally abruptly breaks down at the Curie temperature.

It may properly be inquired why it is that a piece of iron with a Curie temperature of 1043°K may readily be demagnetized by appropriate treatment. The reason for this is that the parallel orientation of spin moments in a ferromagnetic substance occurs in certain small volumes called Weiss domains. Within each domain the spin moments are parallel at all temperatures below the Curie temperature. But the orientation of the resultant moment may be quite different in different domains, even though these domains may be adjacent to each other. It may, therefore, occur that the over-all magnetization of any particular specimen may be vanishingly small if the direction of magnetization in different domains is random. These domains may contain only a few atoms, or they may be large enough to observe under low magnification. Techniques are available for making them visible and for determining the direction of magnetization.

In the presence of an applied field the direction of magnetization in the domains tends to turn in the direction of the field. This turning may be coherent within each domain or it may involve growth of properly oriented domains at the expense of those with moments pointed in other directions. This wholesale alignment is the process of magnetization; the limit is reached when orientation is effectively complete.

The magnetization thus obtained, even though the applied field need not be very large, is often called the "saturation" magnetization. It does, however, vary with temperature and, to some degree with field. Some authors refer to it as the "technical" saturation. Following Stoner[6] we may call it a "quasi-saturation," but a more familiar term is "spontaneous" magnetization.* For the spontaneous magnetization we shall use the symbol I_{sp}.

The spontaneous magnetization of iron at room temperature is 1707 gauss; at 0°K it is 1752 gauss. The variation with temperature, which is much the same for many ferromagnetics, is shown in Fig. 8, in which relative spontaneous magnetization is plotted against T/T_c. The true saturation magnetization is the magnetization at infinite field and absolute zero. This will be designated by M_0. Complete alignment of atomic magnetic dipoles at any temperature other than absolute zero may be achieved, at least in principle, at infinite field. This will produce the saturation magnetization designated M_s.

It will be recalled that the magnetization is the moment per unit volume. The "saturation" moment per atom, $\bar{\mu}_A \beta$, may be obtained from the expression:

$$M_0 = N\bar{\mu}_A\beta \tag{2.13}$$

where N, as before, is the number of particles (in this case, the

---

* Some authors restrict the term spontaneous magnetization to the magnetic moment per unit volume within a domain at temperature T and field zero. We shall find that the definition given above will be satisfactory for our purposes.

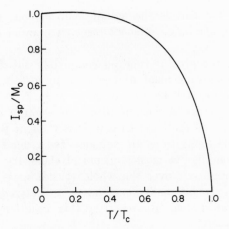

FIG. 8. Relative spontaneous magnetization versus reduced Curie temperature, T/T_c, for a typical ferromagnet.

number of atoms) in a sample of unit volume and β is the Bohr magneton. The saturation moment per atom of iron, expressed in Bohr magnetons, may then be calculated as follows:

$$\tilde{\mu}_{\text{Fe}} = \frac{1.752 \times 10^3 \text{ gauss} \times 5.585 \times 10 \text{ g mole}^{-1}}{7.895 \text{ g cm}^{-3} \times 6.025 \times 10^{23} \text{ mole}^{-1} \times 9.27 \times 10^{-21} \text{ gauss cm}^3}$$

$$= 2.22$$

Some moments and Curie temperatures are given in Table III.

TABLE III

SATURATION MOMENTS IN BOHR MAGNETONS AND CURIE TEMPERATURES
FOR SOME FERROMAGNETICS

Substance	$\tilde{\mu}_A$	T_c °K
Fe	2.221	1043
Co	1.726	1400
Ni	0.606	631

The saturation moment so obtained should not be confused with the paramagnetic moment discussed in a previous section. The paramagnetic moment, expressed in Bohr magnetons, is given by $g[J(J+1)]^{1/2}$ or, if we may consider electron spins only, by $2[S(S+1)]^{1/2}$; whereas, under these conditions, the saturation moment is simply $2S$. The reason for this difference is that the paramagnetic moment is the actual moment, but the saturation moment is the maximum moment resolved parallel to the applied field.

There is at the present time no completely satisfactory theory concerning the arrangement of electrons in metals. A few paragraphs will be inserted here in an effort to make intelligible some of our later remarks. A current theory of electron distribution in metals is known as the band theory. It is thought that the energy levels available to electrons are not localized around atoms or ions in the sense familiar to most students of chemistry but that they are rather spread out over the whole crystal mass. The isolated atoms of iron, cobalt, and nickel have, respectively, 6, 7, and 8 electrons in the d shell. Representing this situation in a familiar manner we have:

Fe $3d^6$ (↑↓) (↑) (↑) (↑) (↑) $4s^2$ (↑↓)

Co $3d^7$ (↑↓) (↑↓) (↑) (↑) (↑) $4s^2$ (↑↓)

Ni $3d^8$ (↑↓) (↑↓) (↑↓) (↑) (↑) $4s^2$ (↑↓)

from which the saturation moments might be expected to be 4, 3, and 2 Bohr magnetons. Actually, the moments are 2.22, 1.73, and 0.61. According to the band theory it is assumed that the $3d$ states and the $4s$ states overlap in such a way that, on the average, fractional filling of the several states is possible. To explain, for instance, the saturation moment of 0.61 Bohr magneton for nickel, it is assumed that of the 10 available $3d$ states electrons actually occupy 9.4. Five of the electrons have spins in one direction; 4.4 in the other direction. This leaves 0.6 unopposed electron spins and these are responsible for the magnetic moment. The remaining 0.6 electron is, on the average, to be found in the $4s$ band, but s electrons are thought to make no contribution to the magnetic moment.

The views stated above, very briefly indeed, gain support from studies of saturation magnetization in nickel alloys, of which nickel-copper is a good example. As alloying proceeds the saturation moment steadily drops, becoming zero at approximately 53 atom % of copper. This is thought to occur from s (valency) electrons from the copper entering the lower lying d band of the nickel and thus filling up the partially empty d subband. The situation is complicated by the probability that there is an orbital contribution to the observed moment of nickel. If this is true then we may be forced to the conclusion that the number of electrons transferred to the nickel per atom of copper is somewhat less than one. These questions have a relation to the interaction of an adsorbed gas on the surface of a ferromagnetic metal. We shall have more to say concerning this matter later.

Some of the ideas given above are represented graphically in Figs. 9 and 10. It must be remembered that some of the electrons in metals contribute to the electrical conductivity, and are said to be in the conduction band.

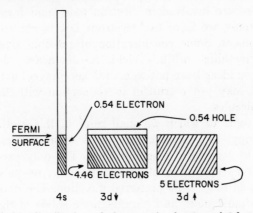

FIG. 9. Probable distribution of electrons in the 4s and 3d states for nickel at absolute zero, according to the band theory (from Kittel[2]).

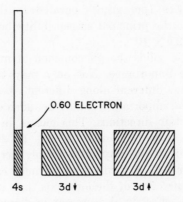

FIG. 10. Probable distribution of electrons in a nickel-copper alloy containing about 53 atom % of copper (from Kittel[2]).

An alternative approach to metals is presented by Pauling.[7] On the basis of physical properties such as hardness, density, and heats of fusion and vaporization, he assigns "valences" to the atoms in metal crystals. These valences are often numerically equal to the maximum oxidation states shown by the elements in their respective compounds and they correspond to the number of electrons used in forming bonds. In the first transition series the valences rise from 1 and 2 for potassium and calcium, respectively, to a maximum of 6 for manganese, iron, cobalt, and nickel, falling to 5½ for copper and progressively lower as we move away from the transition elements. Iron has eight electrons outside the argon shell. If six of these are involved in electron pair bond formation with other iron atoms, we have two electrons left to contribute to the magnetic moment. Some consideration of possible single electron bonds and "metallic" orbitals yields results nearer the observed moment. These ideas have not won wide acceptance, but consideration of them may prove fruitful in connection with the effect of adsorbed molecules.

The final topic to which we shall refer in this section is magnetic anisotropy. In general a sample of crystalline matter tends to orient in a uniform magnetic field. The only exceptions are crystals such as the cubic of high symmetry, or polycrystalline matter in which there is no preferred direction. The reason for this behavior in diamagnetic and paramagnetic solids is that the susceptibility of the molecule may be different in different directions. Crystalline benzene is an example in which the two principal susceptibilities (per gram) parallel to the ring are about -0.5×10^{-6} while the principal susceptibility perpendicular to the ring is about -2.5×10^{-6}.

In ferromagnetic solids the phenomenon of magnetic anisotropy may be of prime importance. Not only may the saturation moments be somewhat different along different axes but, of considerably more practical importance, it may be much easier to magnetize the crystal in certain directions. This means, in other words, that the permeability may be quite different in different directions. Ferromagnetic anisotropy may arise simply because of the shape of the crystal. A needle-shaped crystal is much easier to magnetize, generally, in the direction of the long axis than it is at right angles to this axis. (This effect is also present in diamagnetics and paramagnetics but it is not nearly so pronounced.) But ferromagnetic anisotropy may also arise from the arrangement of atoms in the crystal. Cobalt is, for instance, easy to magnetize along the hexagonal axis but hard to magnetize at right angles to this axis. This phenomenon is known as magnetocrystalline anisotropy; it apparently arises from the effect of electrostatic fields produced by the atoms themselves on the spin-orbital coupling, and partial quenching of the orbital contribution.

There are several other sources of anisotropy. One of the most important is strain anisotropy resulting, as the name implies, from dislocations produced by mechanical strains in the sample. These strains may arise from mechanical working of the sample, but they often appear to an exaggerated degree in small particles or in thin metal films formed by condensation from the vapor.

We shall have reference to magnetic anisotropy later. One consequence is that anisotropy has an influence during demagnetization just as it does during magnetization. If therefore, an anisotropic substance is first magnetized and then the external field is reduced to zero it will be found that the sample still shows a more or less strong magnetization. This means that the spontaneous magnetization in the domains continues to have a preferred direction which may be overcome only by applying an appropriate field in the reverse direction, or by heating the sample above its Curie temperature. These effects give rise, as is well known, to the phenomena of magnetic hysteresis, residual magnetism, and coercive force.

5. Ferrimagnetism[2]

A fairly large group of substances has the spinel crystal structure, and may be represented by the general formula $MO \cdot Fe_2O_3$. These substances are called ferrites. The metal M may, for example, be Zn^{2+}, Mn^{2+}, or Fe^{2+}. Various combinations of ferrites are possible, part of the $(3+)$ iron may be replaced by other elements, and the oxygen may be replaced by sulfur. Combinations with rare earths, known as rare earth garnets, are also well known. Many such substances exhibit a kind of ferromagnetism. Interest in them has developed rapidly because they combine the properties of high magnetization with quite low electrical conductivity. For this reason the ferrites have applications in high frequency electronic equipment.

Ferrites apparently owe their magnetic properties to a circumstance first described by Néel. It will be recalled that in antiferromagnetism the electron spins are arranged antiparallel and hence cancel each other. In ferromagnetics the spins are arranged parallel and hence reinforce each other. In the ferrites some of the spins are parallel and some antiparallel. In, for instance, Fe_3O_4 which is a spinel (inverse) written better as $Fe^{2+}Fe_2^{3+}O_4$, for each Fe^{3+} in a tetrahedral hole there is one Fe^{3+} and one Fe^{2+} in octahedral holes. The saturation moments are 5 Bohr magnetons for Fe^{3+} and 4 for Fe^{2+}. In each Fe_3O_4 group there are five electron spins in tetrahedral coordination directed antiparallel to five spins in octahedral coordination. This leaves four spins in octahedral coordination able to contribute to the ferromagnetism. The observed moment in Fe_3O_4 is about 4.2 Bohr magnetons. But if the Fe^{2+} ions are progressively replaced by diamagnetic Zn^{2+} ions the moment falls to zero. Néel suggested the name "ferrimagnetism" for this kind of magnetic behavior.

REFERENCES

1. L. F. Bates, "Modern Magnetism." Cambridge Univ. Press, New York and London, 1951.
2. C. Kittel, "Introduction to Solid State Physics," 2nd ed., pp. 207, 234, and 402. Wiley, New York, 1956.
3. P. W. Selwood, "Magnetochemistry," 2nd ed., pp. 337 and 380. Interscience, New York, 1956.
4. G. Foëx, in "Constantes Sélectionnées," Vol. 7, Diamagnétisme et Paramagnétisme. Masson, Paris, 1957.
5. R. M. Bozorth, "Ferromagnetism." Van Nostrand, New York, 1951.
6. E. C. Stoner, "Magnetism and Matter," p. 119. Methuen, London, 1934.
7. L. Pauling, "The Nature of the Chemical Bond," 3rd ed., p. 394. Cornell Univ. Press, Ithaca, New York, 1960.

CHAPTER III

Very Small Ferromagnetic Particles

1. Introduction

The presence of a single layer of adsorbed molecules on the surface of a metal could hardly be expected to cause a measurable change in the magnetization of the metal unless the sample of adsorbent had a large specific surface area. With available methods for measuring magnetizations it might be expected that such an effect could be observed if 1% of the metal atoms were on the surface. Reasonable precision might be expected if 10% were so situated.

This requirement means that the metal particles must be rather less than 100 A in diameter and must, therefore, contain no more than a few thousand atoms. It is a fortunate circumstance that the nickel particles in a typical nickel-silica hydrogenation catalyst average about 50 A in diameter, or even less. Commercial nickel catalysts are found, not infrequently, to be quite suitable for magnetic investigation of chemisorption processes of the kind to be described.

A very small particle of a ferromagnetic is essentially a single magnetic domain. Such particles exhibit magnetic properties which are unique, and which lie on the borderline between ferromagnetism and paramagnetism. Michel[1] showed in 1937, and more specifically[2] in 1950, that the slow reduction of nickel-silica catalysts, or of certain other active preparations, may yield nickel in a form which has certain aspects of ferromagnetism but which appears to show no definite Curie temperature. In such preparations the magnetization merely falls, more or less regularly, with increasing temperature. The magnetization generally becomes negligibly small at least 100° below the normal Curie point. Similar

35

behavior was noted[3] on the part of the nickel in a reduced catalyst derived from nickel ammonium molybdate.

Michel's interpretation of this anomaly—an interpretation later assumed and developed somewhat by the author[4]—was that very small particles of ferromagnetic substances should have Curie points lower than that of the same substance in massive form,* and that typical catalyst preparations containing a wide distribution of particle diameters would, therefore, possess a wide range of Curie points, of which none would be well defined. The lowering of the Curie points was thought to be related to the fact that many atoms are on the surface of such particles, and such atoms lack complete coordination. Alternatively, it was thought that such small particles might be imperfectly crystallized and that this effect might also be responsible for the subnormal Curie points.

This interpretation is almost certainly of minor significance insofar as it concerns the magnetization-temperature curves of reduced nickel-silica preparations. But Michel's ideas guided some of the early work on the subject and even now it is not completely certain whether or not particle size has any effect on the Curie temperature. We shall return to this point later.

The view that a particle of a ferromagnetic substance, below a certain critical size, would consist of a single domain was first suggested by Frenkel and Dorfman.[5] The term "single domain" may have several meanings. We shall use it to mean a particle which is in a uniform state of magnetization at any external field. Such a particle may have a diameter of 300 A or less depending on the particular substance. These particles may exhibit a kind of magnetic Brownian movement in such a way that orientation of the magnetic moment, of the particle considered as a whole, is hindered by thermal agitation. The particle, when placed in an external field, tends to behave like a paramagnetic atom, but one which has a very large magnetic moment. That such behavior actually occurs was shown by Elmore[6] who studied colloidal suspensions of magnetite and of *gamma*-ferric oxide. But there is also a mechanism of thermal relaxation not involving physical rotation of the particle.[7]

Several names have been suggested for the magnetic behavior

---

* The word "massive" will be used to mean matter in the form of relatively large, well-crystallized pieces. Some authors use the term "bulk" to mean the same thing, and refer to "bulk" metal.

of a ferromagnetic substance in such a state of subdivision that the magnetization is dependent on field and temperature in a manner similar to that of paramagnetics. One such name is "superparamagnetism" suggested by Bean[8]; others include "apparent paramagnetism," "collective paramagnetism," "quasi-paramagnetism," "quasi-ferromagnetism," and "subdomain behavior." The only objection we have to "superparamagnetism" is that the name was used previously[9] for another effect which seems, at this writing, to have been discredited.* This previous use has led to confusion.

2. Collective Paramagnetism†

The first treatment of collective paramagnetism appears to be that of Stoner,[11] which was based in part on earlier considerations of Gans and Debye.[12] Stoner's suggestion was made in an attempt to explain the peculiarly large values of $\delta M/\delta H$ observed for pure, massive nickel. He pointed out that the anomaly could be accounted for by assuming the presence of single domain particles containing not more than a few thousand atoms. The magnetization of an assembly of such particles would be described by the Langevin equation (2.4), except that the moment, μ, of such a particle is equal to $\bar{\mu}_A n$ where n is the number of atoms in the particle. We note also that as the magnetization is the moment per unit volume, we may write:

$$\mu = I_{sp}v \tag{3.1}$$

where v is the volume of a particle. Hence, for an assembly of uniform particles,‡

* A review, and defense, of the effect previously called "superparamagnetism" (sometimes translated as "hyperparamagnetism") will be found in an article by N. I. Kobozev et al. [J. Phys. Chem. U.S.S.R. (English translation) 33, 641 (1959)]. Certain aspects of the work reported by Kobozev appear to have anticipated later developments in the magnetic properties of very small particles, but various other claims have not been confirmed.

† This and the following three sections are based primarily on the papers of C. P. Bean and his associates at the General Electric Research Laboratories in Schenectady.[10]

‡ It might be thought that, as is the case for paramagnetics, it would be necessary to use the Brillouin,[13] rather than the Langevin function to describe the magnetization at very low temperatures. But a particle containing several hundred atoms may be thought of as having a spin quantum number, S, in the hundreds. Such a particle is adequately described by classical theory. It will be noted also that the difference between paramagnetic and satura-

$$M = I_{sp}V\left(\coth\frac{I_{sp}vH}{kT} - \frac{kT}{I_{sp}vH}\right) \tag{3.2}$$

V being the sum of volumes of all particles in a sample.

For all real adsorbent samples to be discussed here, the particle volume is far from uniform. Let us assume a distribution of particle volumes, $f(v)$, where:

$$\int_0^\infty f(v)dv = 1 \tag{3.3}$$

Then[14] the magnetization of such an assembly is:

$$M = V\int_0^\infty I_{sp}\left[\coth\left(\frac{I_{sp}vH}{kT}\right) - \frac{kT}{I_{sp}vH}\right]f(v)dv \tag{3.4}$$

If I_{sp} is independent of v:

$$M = VI_{sp}\int_0^\infty\left[\coth\left(\frac{I_{sp}vH}{kT}\right) - \frac{kT}{I_{sp}vH}\right]f(v)dv \tag{3.5}$$

This equation is in a form which permits correction for the variation of I_{sp} with temperature. It will be noted that an average particle volume may be obtained from the variation of M with H. This useful procedure will be discussed in detail in a later section.

The treatment given above is based on the assumption that no remanence is ever present, or, to express it another way, that thermal equilibrium is established in a time short compared with the duration of the measurements. This assumption is often justifiable at room temperature and above, but it is never justifiable in any catalyst sample thus far studied at very low temperature. The reasons for this have an important bearing on our over-all problem. They will be discussed in the next section.

3. Anisotropy Effects

Certain geological deposits of ferromagnetic iron oxide are found to be magnetized in a direction not readily to be related to the present direction of the earth's magnetic field. As part of an attempt to explain this anomaly Néel[7] developed a theory of magnetization in small particles. This was based partly on the ideas previously developed by Stoner. Further developments and ap-

---

tion moments varnishes. This difference, which is the difference between $2[S(S+1)]^{1/2}$ and $2S$, becomes negligible as S becomes quite large.

plications have been made by C. P. Bean and his associates at the General Electric Research Laboratories.* Real particles are never truly isotropic. Let there be a particle of moment μ directed at an angle θ to an applied field, H. If the anisotropy of the particle is uniaxial, the anisotropy contribution to the total energy may be:

$$E_K = Kv \sin^2 \theta \tag{3.6}$$

where K is the anisotropy energy per unit volume, and where the Boltzmann distribution of the angles θ to the field will be different than if the particle were isotropic.

As shown previously, anisotropy may arise from various sources. In general the anisotropy energy is proportional to the volume of the particle. Large particles, or elongated particles, may deviate substantially from behavior analogous to true paramagnetism or, in other words, they may no longer be described in terms of collective paramagnetism. Such an assembly of particles may be magnetized, but if the external field is removed, the magnetization will be lost in a finite time which may be very long indeed. Néel showed that decay of the remanent magnetization, M_r, proceeds exponentially as follows:

$$M_r = M_s \exp\left(-t/\tau\right) \tag{3.7}$$

where:

$$1/\tau = f_0 \exp\left(-Kv/kT\right) \tag{3.8}$$

where t is the time, τ is the relaxation time, and f_0 is a frequency factor of the order of 10^9 sec^{-1}. Random orientation after removal of the external field thus requires an activation energy. One solution to Néel's original problem concerning the ferromagnetic rocks is that the deposits were indeed laid down, in ancient times, parallel to the earth's field, but that the position of the North Pole has changed before the particles have had time to revert to random orientation. The relaxation time for these rock particles is, therefore, millions of years.

Some idea of particle volumes and relaxation may be found from Eqs. (3.7) and (3.8) with the aid of the anisotropy constants given by Bozorth[15] (see Table IV). For our purposes, we shall

---

* It is remarkable that a development in geophysics plus one in the precipitation hardening of metals should have applications in heterogeneous catalysis.

TABLE IV
MAGNETIC ANISOTROPY CONSTANTS (ERG CM^{-3}) FOR
IRON, COBALT, AND NICKEL

Temp. °K	Fe (bcc) $K_1 \times 10^{-3}$	Co (hcp) $(K_1 + K_2) \times 10^{-6}$	Ni (fcc) $K_1 \times 10^{-3}$
4.2	575	9.5	−750
77	560	9.5	−650
300	480	4	−35

simply point out that Kv is a measure of the energy barrier over which the direction of magnetization in the particles has to be reversed by thermal activation.

To simplify the calculation the method followed will be that of Bean and Livingston.[10] This is to consider a relaxation time of $\tau = 10^2$ seconds to be a criterion of collective paramagnetism; and that the energy barriers along certain crystallographic axes have the following values: $Kv/4$ for $K > 0$, ([100] easy direction); and $Kv/12$ for $K < 0$, ([111] easy direction). Cobalt is a rather special case for which the barrier is taken as $(K_1 + K_2)v$.*

If $\tau = 10^2$ seconds, then $v \simeq 25 \ kT/K_1$ from which the radii given in Table V may be found. These are the radii of spherical particles, calculated with certain simplifying assumptions, for which the saturation magnetization will decay to 1% in about 6 or 7 minutes. If the particles are not spherical the time for decay will be longer.

TABLE V[a]
CRITICAL RADII[b] FOR DECAY OF M_s TO M_r

Metal	Critical radii (A)		
	300°K	77°K	4.2°K
Fe (bcc)	127	77	29
Co (hcp)	48	19	7
Ni (fcc)	438	105	38

[a] The writer is indebted to Dr. Carlos Abeledo for making these calculations.
[b] Based on relaxation time, $\tau = 10$ seconds.

* The reason for using $K_1 + K_2$ for cobalt is discussed briefly by Kittel (ref. 13, p. 429).

While the radii given in Table V are approximate only, they give some idea of the magnitudes, and they show that it requires a smaller particle of cobalt than of nickel to exhibit collective paramagnetism. It will be noted that the rate at which M_s decays to M_r is quite sensitive to particle radius.

4. Some Experimental Evidence for Collective Paramagnetism

Evidence of collective paramagnetism was presented by Heukelom et al.[16] in 1954. These authors measured the magnetization of reduced nickel-silica catalyst preparations over a range of field strength and at two temperatures, 80° and 300°K, as shown in Fig. 11. One characteristic property of matter exhibiting collec-

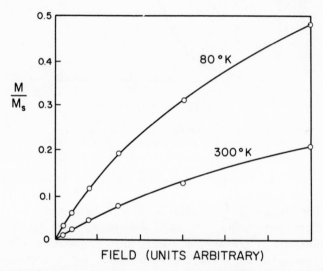

FIG. 11. Relative magnetization as a function of field at two temperatures. The sample was a reduced nickel-silica catalyst which clearly shows the predicted relation of M to H/T (after Heukelom, Broeder, and van Reijen).

tive paramagnetism is that magnetizations obtained at different temperatures may be superimposed if plotted with respect to H/T. The behavior is, of course, simply a consequence of Eq. (3.2). A correction must be made for the change of I_{sp} with T. The data of Heukelom et al. clearly show that this superposition occurs.

Another example of collective paramagnetism is given by

Becker[14] who investigated small particles of cobalt in copper. The procedure was to quench a 2% cobalt solid solution from 1050°C. The alloy was then heated briefly at 650° and quenched again. This procedure caused precipitation of cobalt particles averaging only 12 A in radius, and growing perceptibly with time as they aged. Figure 12 shows the H/T superposition curve as obtained at

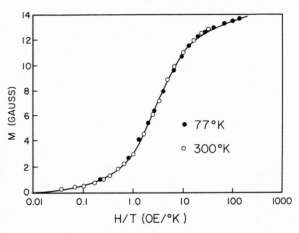

Fig. 12. Collective paramagnetism (superparamagnetism) shown by precipitated cobalt particles in a copper matrix (after Becker).

77° and 300°K. The agreement with prediction is quite satisfactory.

Similar studies of cobalt and of iron particles have been made by various investigators. We shall describe in some detail the results obtained by Dietz[17,18] on nickel-silica preparations. Figure 13 shows magnetizations (in arbitrary units) for a coprecipitated nickel-silica prepared according to the method of van Eijk van Voorthuysen and Franzen[19] (CLA-5421) and containing about 32% of nickel by weight. The sample was reduced in flowing hydrogen for 12 hours at 350°C and then evacuated to about 10^{-6} mm Hg over a period of hours at the same temperature. Figure 14 shows the same data plotted as a function of H/T, after appropriate corrections for demagnetizing fields (see p. 57) and for the change of spontaneous magnetization with temperature. The superposition is good although perhaps not quite as good as might have been hoped. The effect of magnetic anisotropy in somewhat

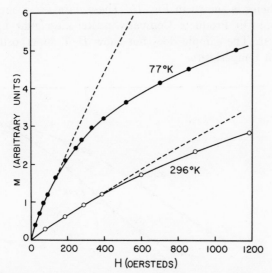

FIG. 13. Field strength dependence of magnetization shown by a reduced nickel-silica catalyst prepared by the coprecipitation method (after Dietz).

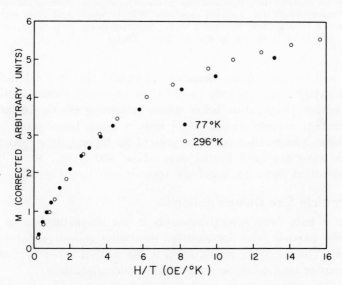

FIG. 14. The data of Fig. 13 replotted to show magnetization as a function of H/T with appropriate corrections for demagnetizing fields and the change of I_{sp} with temperature (after Dietz).

larger particles is shown in Fig. 15. These data are from a sample of Universal Oil Products Company nickel-kieselguhr hydrogenation catalyst. The sample does not show H/T superposition much below room temperature.

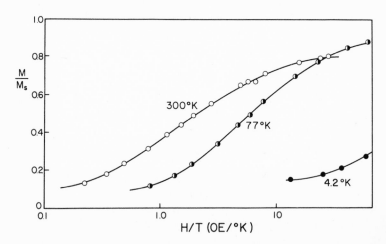

FIG. 15. Effects due, presumably, to anisotropy in preventing M versus H/T superposition for a nickel-kieselguhr sample which had been sintered and which consequently contained nickel particles larger than those used in obtaining the data shown in Fig. 14 (after Dietz).

It will be clear from these several results that typical nickel-silica catalyst samples may indeed show collective paramagnetism. The critical temperature below which anisotropy effects (including remanence) become serious is not usually much below room temperature. The smaller particles present in coprecipitated samples which have not been heated over about 400°C will show H/T superposition down to liquid air temperature (see Fig. 16).

5. Particle Size Determination

There have been several attempts to use magnetic data to determine particle sizes for samples exhibiting collective paramagnetism. The following presentation is the method given by Cahn[20] as adapted by Dietz[17] to the systems under discussion.

In principle the magnetic data at high field provide a more reliable indication of particle size distribution, but it is at high field (and low temperature) that the effects of anisotropy become im-

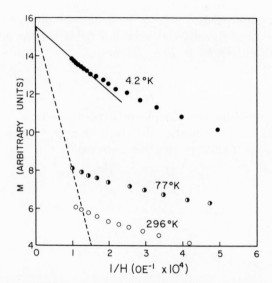

FIG. 16. Magnetization versus reciprocal field for a reduced nickel-silica sample at several temperatures. This shows how important it is to obtain data at low temperatures if a meaningful extrapolation to the saturation magnetization at $0°K$ is required. The dashed line shown was used to estimate an upper limit for the average particle volume, $\bar{v}$, from the high-field data (after Dietz).

portant. We shall start with the low-field approach. For uniform particles [see Eq. (2.5)],

$$M = I_{sp}V \frac{I_{sp}vH}{3kT} \tag{3.9}$$

where, as before, v is the volume of a particle and V is the total volume of ferromagnetic particles in a sample. Writing $V = v_i n_i$ where n_i is the number of particles of volume v_i in the sample, we have:

$$M = \frac{I_{sp}^2 H}{3kT} \sum_i n_i v_i^2 \tag{3.10}$$

For all the measurements to be described later, M_s is approximately equal to $I_{sp} \Sigma_i n_i v_i$; hence, as it is convenient to use relative magnetizations:

$$\frac{M}{M_s} = \frac{I_{sp}H}{3kT} \frac{\Sigma_i n_i v_i^2}{\Sigma_i n_i v_i} \tag{3.11}$$

and the average particle volume $\bar{v}^2/\bar{v}$ may be obtained from the initial slope (i.e., the slope at very low fields) of the curve of M plotted with respect to H/T, as follows:

$$\frac{\bar{v}^2}{\bar{v}} = \frac{3kT}{I_{\mathrm{sp}}H} \left(\frac{M}{M_s}\right) \tag{3.12}$$

Before presenting an example of average particle volume calculation we shall consider the high-field approximation. For $M/M_s \simeq 1$, the Langevin function becomes:

$$\frac{M}{M_s} = 1 - \frac{kT}{\mu H} \tag{3.13}$$

which, for the situation under consideration, gives:

$$\frac{M}{M_s} = 1 - \frac{kT\Sigma_i n_i}{I_{\mathrm{sp}}H\Sigma_i n_i v_i} \tag{3.14}$$

and this yields

$$\bar{v} = \frac{kT}{I_{\mathrm{sp}}H} \cdot \frac{1}{1 - (M/M_s)} \tag{3.15}$$

Inspection of Fig. 16 will show the futility of trying to obtain a value for M_s at any temperature much above 4.2°K on samples of this kind. Measurements are, therefore, first made on the sample at the boiling point of helium or, preferably, even lower. A plot of magnetization (or of galvanometer deflection) versus reciprocal field will then give a value corresponding to M_s. Thus, in Fig. 16, M_s is 15.6 (in arbitrary units). The sample is then warmed to conditions under which true collective paramagnetism is exhibited as evidenced by M vs. H/T superposition as previously described. For the sample for which data are shown in Fig. 16 this occurs at, and above, 77°K. The initial slope of the magnetization as a function of field is found from Fig. 13. Taking, say, the slope of the initial points at 296°K, we find a deflection of 3.1 at 1000 oersteds. This gives $M/M_s = 3.1/15.6 = 0.199$. Then, assuming that the spontaneous magnetization of very small particles of nickel is the same as that of massive nickel,* namely, 485 gauss at 296°K, we have:

$$\bar{v}^2/\bar{v} = \frac{3 \times 1.38 \times 10^{-16} \text{ erg deg}^{-1} \times 2.96 \times 10^2 \text{ deg} \times 1.99 \times 10^{-1}}{4.85 \times 10^2 \text{ gauss} \times 10 \text{ oersted}^3}$$

$$= 50 \times 10^{-21} \text{ cm}^3$$

* Some further reference to this assumption will be made later (p. 71).

The dashed line drawn in Fig. 16 represents an attempt to find the "high-field" magnetization slope so that we may use Eq. (3.15) to calculate $\bar{v}$. The average volume, $\bar{v}$, so found is 4.5×10^{-21} cm³.

Spherical particles of nickel with a volume of 50×10^{-21} cm³ would have a specific surface of 156 m². These particles are too small to be studied by X-ray linewidth broadening with any degree of precision. The sorption of hydrogen at room temperature and 1 atm of pressure is about 24 cm³ g^{-1} of nickel. If one may make the doubtful assumption that each hydrogen atom covers 6.4 Å², we may estimate the specific surface as about 43 m². The chemisorption of carbon monoxide gives areas of between 31 and 75 m² for similar samples. These results are about as satisfactory as could be expected in view of the fact that the average v must always be smaller than $\bar{v}^2/\bar{v}$.

There have been several attempts to estimate particle sizes, of systems such as these, without the rather tedious and direct determination of the saturation moments. Heukelom et al.[16] found that the magnetizations as obtained over a range of field strength at room temperature could be expressed by the empirical relation

$$\frac{\sigma}{\sigma_\infty} = \frac{(\alpha H)^{0.9}}{1 + (\alpha H)^{0.9}} \tag{3.16}$$

where σ is the specific magnetization, σ_∞ is the specific saturation magnetization at 0°K, and α is a constant. From this relation a value for σ_∞ and hence for the effective moment, μ of the nickel particle may be obtained by a lengthy extrapolation.

A refinement on this method has been reported* by Trzebiatowski and Romanowski.[21] By successive approximations the complete Langevin equation is made to yield a value for σ_∞. The procedure in both of these approaches is then to divide the effective moment, μ, by the moment, $\bar{\mu}_{Ni}$, for a nickel atom in massive nickel. The particle diameters so obtained are comparable with those given by the more accurate method described in detail above; nevertheless, these methods are essentially based on the low field magnetizations. Discussion of the interesting conclusions reached by these several authors will, therefore, be deferred until we have completed our presentation of the method and results involved in the experimental approach to the saturation magnetization. The general area of

---

* It is a matter of regret to the author that the interesting paper of Trzebiatowski and Romanowski did not come to his attention until nearly 4 years after it was published.

magnetic particle size determination by application of the Langevin equation is discussed by Bean and Jacobs,[22] and by Knappwost.[23] In 1954 the author proposed a method[4] based on the assumption that the Curie temperature is a function of particle size. This assumption has since been shown to be wrong, or at least to be of minor significance in connection with the systems under investigation. A useful result which emerged from these several studies[16] is that typical nickel-silica preparations begin to sinter, with growth of the nickel particles, if they are heated above about 400°C.

In a rather different category is the method proposed by Weil.[24-26] This method is based on the remanence shown by very small ferromagnetic particles at low temperature. The method is to measure the ratio of remanence to saturation, M_r/M_s, as a function of temperature. The remanence observed at a given temperature shows the fraction of the mass with particle volumes greater

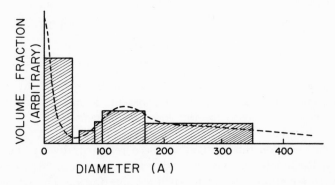

FIG. 17. The apparent distribution of particle volumes in a Raney nickel catalyst sample (after Weil).

than that just capable of behaving according to the Langevin function. In practice it is convenient to find the time interval during which the remanence decays to a fixed fraction of M_s or, alternatively, the ratio M_r/M_s after decay has proceeded for a fixed number of seconds. Weil has applied this method to Raney nickel* with results such as those shown in Fig. 17. Assuming that the particles are approximately spherical one may calculate the sur-

---

* Raney nickel is always of interest to specialists in catalysis, but it is an impossibly complicated mixture on which to do studies of this kind.

face area corresponding to the particle size distribution found magnetically. The results given by Weil are 104 and 85 m² g⁻¹ for two samples for which the BET surface area is 100 m² g⁻¹. This agreement must in part at least be fortuitous because no consideration was given to the effect of the paraffin or alcohol which were used to suspend the nickel. As we shall see later any adsorbed molecule causes a substantial loss of magnetization in such samples.

It is possible to draw some conclusions concerning particle shape from the method proposed by Weil, as mentioned above. Another source of information about particle shape has been suggested by Bean et al.[27] Ferromagnetic resonance absorption depends, to some degree, upon the shape of the specimen. It is possible to distinguish between spheres, platelets, and rods, and this has been done for small particles of precipitated cobalt.

The method appears to be applicable to systems of interest in heterogeneous catalysis. For instance, nickel supported on silica gel shows ferromagnetic resonance absorption[28] (at $g = 2.2$) the symmetry of which suggests spherical nickel particles.

It may, in conclusion, be pointed out that collective paramagnetism may be observed in substances other than metals. It was mentioned above that Elmore[6] had noted some of the effects in colloidal magnetite and in gamma-ferric oxide. The latter has recently been investigated in more detail by Knappwost and Stein.[29] Gamma-ferric oxide may be made in reasonably well-controlled particle diameters by the method of Haul and Schoon[30] in which iron pentacarbonyl vapor is carried in a stream of nitrogen until it meets air or oxygen at a temperature between 280° and 600°C. The oxidation product is relatively pure and it is stable in air. This interesting development appears to make possible the extension to at least one transition metal oxide of the methods for studying chemisorption to be described in detail in the following chapters.

REFERENCES

1. A. Michel, Ann. chim. 8, 317 (1937).
2. A. Michel, R. Bernier, and G. LeClerc, J. chim. phys. 47, 269 (1950).
3. H. Morris and P. W. Selwood, J. Am. Chem. Soc. 65, 2251 (1943).
4. P. W. Selwood, S. Adler, and T. R. Phillips, J. Am. Chem. Soc. 77, 1462 (1955).
5. J. Frenkel and J. Dorfman, Nature 126, 274 (1930).

6. W. C. Elmore, *Phys. Rev.* **54**, 1092 (1938).
7. L. Néel, *Ann. géophys.* **5**, 99 (1949).
8. C. P. Bean, *J. Appl. Phys.* **26**, 1381 (1955).
9. N. I. Kobozev, V. B. Evdokimov, I. A. Zubovoch, and A. N. Mal'tsev, *Zhur. Fiz. Khim.* **26**, 1349 (1952).
10. C. P. Bean and J. D. Livingston, *J. Appl. Phys.* **30**, 120S (1959).
11. E. C. Stoner, *Phil. Trans. Roy. Soc. London Ser. A*, **235**, 165 (1936).
12. R. Gans and P. Debye, "Handbuch der Radiologie" (E. Marx, ed.), Vol. VI, p. 719. Leipzig, 1925.
13. C. Kittel, "Introduction to Solid State Physics," 2nd ed., p. 216. Wiley, New York, 1960.
14. J. J. Becker, *Trans. Am. Inst. Mining, Met. Petrol. Engrs.* **209**, 59 (1957).
15. R. M. Bozorth, "Ferromagnetism," p. 567–568. Van Nostrand, New York, 1951.
16. W. Heukelom, J. J. Broeder, and L. L. van Reijen, *J. chim. phys.* **51**, 474 (1954).
17. R. E. Dietz, Thesis, Northwestern University, Evanston, Illinois, 1960, p. 68.
18. R. E. Dietz and P. W. Selwood, *J. Chem. Phys.* **35**, 270 (1961).
19. J. J. B. van Eijk van Voorthuysen and P. Franzen, *Rec. trav. chim.* **70**, 793 (1951).
20. J. W. Cahn, *Trans. Am. Inst. Mining, Met. Petrol. Engrs.* **209**, 1309 (1959).
21. W. Trzebiatowski and W. Romanowski, *Roczniki Chem.* **31**, 1123 (1957).
22. C. P. Bean and I. S. Jacobs, *J. Appl. Phys.* **27**, 1448 (1956).
23. A. Knappwost, *Z. Elektrochem.* **61**, 1328 (1957); **63**, 278 (1959).
24. L. Weil, *J. chim. phys.* **51**, 715 (1954).
25. L. Weil, L. Gruner, and A. Deschamps, *Compt. rend. acad. sci.* **244**, 2143 (1957).
26. W. Henning and E. Vogt, *Z. Naturforsch.* **12a**, 754 (1957).
27. C. P. Bean, J. D. Livingston, and D. S. Rodbell, *Acta Met.* **5**, 682 (1957).
28. D. P. Hollis and P. W. Selwood, *J. Chem. Phys.* **35**, 378 (1961).
29. A. Knappwost and H. D. Stein, *Z. Elektrochem.* **64**, 321 (1960); also C. J. Lin, *J. Appl. Phys.* **32**, 233S (1961).
30. R. Haul and T. Schoon, *Z. Elektrochem.* **45**, 663 (1939).

CHAPTER IV

The Measurement of Saturation Magnetization

1. The Experimental Problem

A major purpose of this work is to improve our understanding of the binding forces operative between adsorbent and adsorbate. One method of doing so, in appropriate cases, is to measure the change in the magnetic moment of the adsorbent atoms per molecule of vapor adsorbed. We shall now describe the experimental arrangement for these measurements.

If one deals with a ferromagnetic sample there can be no question that accurate determination of the magnetic moment per atom, $\bar{\mu}_A$, requires a measurement of M_s at temperatures sufficiently low that an extrapolation to find M_0 is feasible. But for true paramagnetic samples it suffices to measure the susceptibility over a moderate range of temperature so that one may calculate μ_A, and see how μ_A may change with changing coordination or changing oxidation state. It might, therefore, be thought that samples exhibiting collective paramagnetism would offer no particular problem; but this is not so. In a true paramagnetic all the (atomic) particles have the same magnetic moment. But in all real collective paramagnetics the particle size, and hence the moment $\mu_A n$, is far from uniform. Furthermore, collective paramagnetic moments depend in a rather peculiar way on the spontaneous magnetization of the atoms within each particle. We have no *a priori* assurance that this spontaneous magnetization is the same as that in massive metal.

From the start of this work it has, therefore, been obvious that some method would have to be devised for obtaining the saturation magnetization, M_0, either directly or indirectly. Two attempts to do this were mentioned in the previous chapter. These were the method of Broeder *et al.*[1] who relied on a very long extrapolation

51

to $1/H = 0$, based on actual measurements at room temperature only; and the refinement introduced by Trzebiatowski and Romanowski[2] and later further improved by Romanowski.[3] Attempts to extend measurements down to liquid hydrogen temperature showed, in the writer's laboratory[4,5] that no very great difference exists between M_0 for massive nickel and for nickel in reduced nickel-silica preparations, but these measurements were made at one field only.

Each of these attempts at finding M_0 may be characterized as essentially a "low-field" method in that the largest fractional M/M_0 actually measured was rather small. The difficulty in obtaining M_0 is then, as pointed out by Bean and Jacobs,[6] inherent in the nature of collective paramagnetism. In a plot of M vs. H/T (as in Fig. 12) the slope at low values of M/M_s is determined more by the larger, more readily magnetizable, particles; while the slope nearing saturation is determined more by smaller particles. An extrapolation based on low fractional saturations cannot, therefore, give better than a rough estimate of M_0 and hence of $\bar{\mu}_A$.

The above remarks should not be construed as meaning that low saturation studies are of no value. On the contrary, we shall see later that for convenience, and for providing useful information, low saturation studies are almost unparalleled. But an accurate estimate of $\bar{\mu}_A$ or of the change in $\bar{\mu}_A$ produced by an adsorbed molecule cannot be made in this way.

Our purpose requires that magnetizations should be measured at fields high enough and at temperatures low enough so that extrapolations to find M_0 may be made with confidence. The samples to be studied are pyrophoric in air and hence must be reduced and subsequently handled in closed containers. We shall require a quantitative determination of the amount of adsorbent present. If this adsorbent is a metal it will probably be necessary to obtain the amount actually present as reduced metal and not as oxide or other nonadsorbing form. It will also be necessary to make a quantitative measurement of the amount of vapor adsorbed. Some complications are introduced by the necessity that during measurements of magnetization the sample must be at liquid helium temperature or lower. Other requirements are that the sample should be in a uniform field during measurement, and that provision should be made for carrying out chemical treatment, such as reduction in flowing hydrogen at 400°C, evacuation, and so forth, all *in situ*. The most appropriate method for these purposes appears to be that of Weiss and Forrer.

2. The Method of Weiss and Forrer[7]

The following description will be based on the adaptation developed by Dietz[8] for the study of nickel-silica preparations, and for handling and measurement of hydrogen in contact with the sample. But the description applies specifically to an improved apparatus in which use is made of a 12 inch (pole diameter) magnet with adequate power supply and control.

The field is produced by a Pacific Electric Motor Company Model 12A-HI water-cooled electromagnet shown in Fig. 18. Power

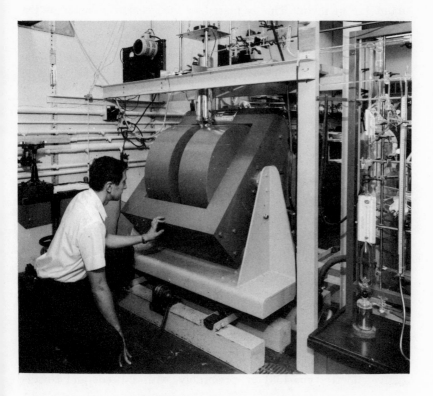

FIG. 18. Magnet assembly for measurement of saturation magnetization at low temperatures.

is obtained at up to 2200 volts from a stabilized source, Model RC-2-2300. The stability of the field has proved to be more than adequate over long periods of time. The pole gap is 2¾ inches, and was produced between truncated conical tips brought down from 12 inches to 4 inches face. The field is uniform to within a small fraction of 1% over a volume of at least 1 cubic inch between the poles. The field is variable from about 18 kgauss down to zero; and it may be reversed as needed. The magnet is mounted on tracks to permit lateral displacement up to 6 feet. The field is monitored with a Rawson Model 720 rotating coil gaussmeter. While the Rawson meter gives the field to ±0.5%, reproducibility of the field by control of the magnet current is, at very least, accurate to 0.01%.

The sample consists of one pressed cylindrical pellet containing about 1 g of metal. This is placed in the geometric center of the pole gap. Two small Helmholtz sensing coils are placed coaxially to the pole pieces—one on each side of the sample. These coils are about 1 cm in diameter and consist of 1000 turns of No. 40 Ceroc (ceramic-Teflon) insulated copper wire.* The coils are connected in series to a ballistic galvanometer.

When a measurement is to be made the sample is lifted mechanically to a position a short distance above the coils, but still in the region of maximum magnetic field. In this way the lines of force passing through the sample are forced to cut the coils in such a manner as to induce a current in the coils. This current is integrated by the ballistic galvanometer to give a reading which may be shown[8] to be proportional to the magnetization of the sample. The arrangement is shown diagrammatically in Fig. 19.

If the magnet power supply is adequately stabilized the arrangement described above may give sufficient precision. Otherwise it may be necessary to provide two pairs of sensing coils connected as shown in Fig. 20. The sample is then raised from between one pair of coils to a position between the second pair. This arrangement balances out transient changes in the applied field. All the coils are in series, but the upper pair of coils is wound in opposition to the lower. The integrated current induced in these coils may be measured with a General Electric light-beam-fluxmeter (galvanometer type-153) or, preferably, with a Leeds and North-

* Made by Sprague Electric Company, North Adams, Massachusetts.

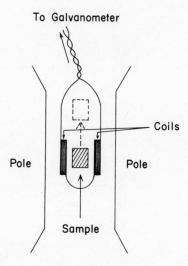

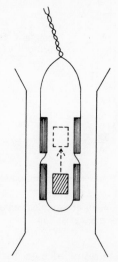

FIG. 19. Relation of sample to sensing coils for measurement of saturation magnetization.

FIG. 20. Arrangement of two pairs of sensing coils to minimize the effect of field fluctuations during the measurement of saturation magnetization.

rup 2290 galvanometer. The method of raising, or lowering, the sample between the sensing coils is shown in Fig. 21.

Measurements at 4.2°K are made with the sample surrounded

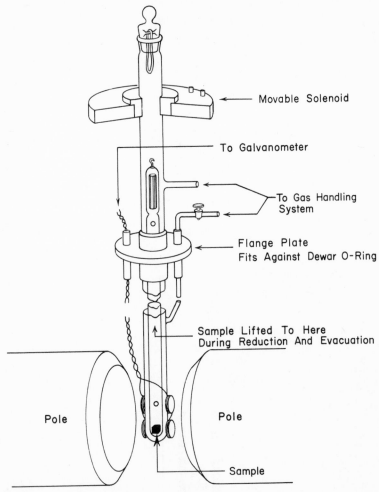

FIG. 21. The general arrangement of sample, container, and magnet poles for measurement of saturation magnetization.

by a Dewar flask of conventional design for this purpose except that the portion of the Dewar between the magnet poles is shielded by a copper screen cooled by, and projecting down from, the liquid nitrogen shield.* Measurements at the *lambda* point of helium, 1.8°K, are made by pumping on the liquid helium in the usual way.

The samples are, of necessity, handled in the absence of air

* The Dewar was made by H. S. Martin and Company, Evanston, Illinois.

after reduction. They are conveniently reduced *in situ* by raising the sample holder to a position high enough to avoid damage to the sensing coils. A small sleeve furnace provides the proper reduction temperature.

The gas handling system is also of conventional design as shown in Fig. 22. In brief, the sample is reduced by flowing purified hy-

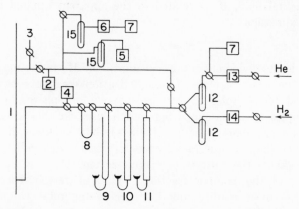

FIG. 22. Gas handling system for saturation magnetization studies: 1, apparatus shown in Fig. 21; 2, cold cathode gauge; 3, hydrogen exhaust; 4, turntable McLeod gauge; 5, McLeod gauge; 6, oil diffusion pump; 7, mechanical pumps; 8, closed arm oil manometer; 9, open arm mercury manometer; 10, gas microburet; 11, gas buret; 12, silica gel traps; 13, helium purification train; 14, hydrogen purifier; 15, traps cooled with liquid nitrogen (after Abeledo).

drogen for many hours. For a typical nickel-silica sample the reduction temperature is 360°C. The sample is then evacuated for a minimum of 2 hours at 360° to a pressure of about 10^{-6} mm Hg, and allowed to cool to the temperature of measurement. A trace of purified helium is added to promote attainment of thermal equilibrium. After the magnetization is measured at whatever temperature is desired, the sample is warmed to room temperature, after which a measured volume of adsorbate gas is admitted. The pressure in the dead-space must, of course, be kept low. The sample is then cooled again for a final measurement of magnetization as affected by a known quantity of adsorbed vapor.

3. Correction for Demagnetizing Field

The field acting to magnetize a sample is always less than the field in the absence of the sample. The reason for this is that the free poles at the ends of the oriented dipoles produce a demagnetizing effect which is dependent on the shape of the sample and on the magnetization. For samples of the kind under consideration the demagnetization correction has been considered by Trzebiatowski,[2] and also by Dietz.[8]

The actual field, H, is related to the apparent applied field H_0 by the expression:

$$H = H_0 - \eta M_T \qquad (4.1)$$

where η is the demagnetization constant, and M_T is the magnetization of the whole sample. Demagnetization constants have been calculated for samples of various shapes; for spheres, $\eta = 4\pi/3$. The samples of silica-supported metals used in this kind of investigation are short cylinders (pellets), but we know little concerning the shape of the metal particles within each pellet. We do, however, know the saturation magnetization, M_s, for the several metals; and the volume fraction, V/V_T, of ferromagnetic substance present is readily found, V_T being the total volume of a sample, including the silica (or other) supporting medium. If we write:

$$\eta M_T = \eta \cdot \frac{M}{M_s} \cdot I_{sp} \cdot (V/V_T) \qquad (4.2)$$

then, for instance, from Fig. 16 it is seen that at room temperature and 5000 oersteds the fraction $M/M_s = 0.34$, whence as $V/V_T \simeq 0.1$, ηM_T must be about 114 oersteds.

One sees, therefore, that the demagnetizing field will not be a negligible fraction of H_0 until saturation is approached at fields of the order of 10^4 oersteds. Figure 23 shows data on a sample of massive nickel obtained at 77°K before and after correction for demagnetization. As expected, the curves converge at high field. All the magnetization data reported here have been corrected, wherever necessary, for the demagnetization factor.

4. Correction for the Magnetic Image Effect

A specimen between the poles of a magnet may induce magnetic charges, or images, in the pole tips. These images have the effect of increasing the *apparent* magnetization of the sample, and the result can be serious. This effect was studied by Weiss and Forrer;

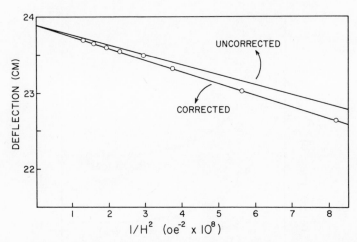

FIG. 23. The approach to saturation magnetization for massive nickel at 77°C, before and after correction for the effects of demagnetizing fields (after Dietz).

to the best of the writer's knowledge the only author to consider this correction as applied to catalytically active solids is Dietz.[8]

It is reasonable to assume that at very large pole gaps the image effect will be negligible. The magnitude of the effect may be demonstrated by measuring the apparent magnetization of a sample at constant field. Results on a coprecipitated nickel-silica sample are shown in Fig. 24 for several fields. It is assumed that the true magnetization is being measured when the apparent magnetization becomes independent of the pole gap.

The apparent magnetization, M_{app} is thus equal to the true magnetization, M, plus an added contribution caused by the image effect. We may write

$$M_{app} = M[1 + f(w,p)] \qquad (4.3)$$

where $f(w,p)$ is a function of the pole gap and of the permeability of the pole tips. Then if M_{app} is measured at two different field strengths, H_1 and H_2,

$$\frac{M_{app}(H_1)}{M_{app}(H_2)} = \frac{M(H_1)[1 + f(w,p_1)]}{M(H_2)[1 + f(w,p_2)]} \qquad (4.4)$$

and

$$\frac{M_{app}(H_1)}{M_{app}(H_2)} = \frac{M(H_1)}{M(H_2)}$$

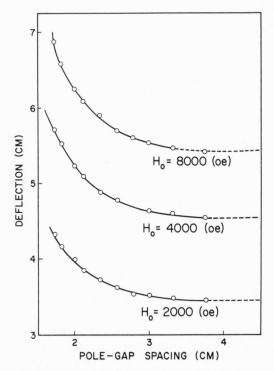

FIG. 24. Apparent magnetization at several pole gap spacings.

provided $f(p_1) \simeq f(p_2)$. That this view is correct may be shown by normalizing the values of apparent magnetizations shown in Fig. 24 by the magnetization at 8000 oersteds corresponding to the same pole gap. When this is done it is found that the normalized values are independent of gap over the range, 1000–8000 oersteds, investigated.

The correction for the image effect may now be found in the following manner.

All specimens investigated show some remanence, although for some samples this does not become measurable until we reach quite low temperatures. We measure M_r' at a given pole gap, but with the field equal to zero as measured by the gaussmeter. The magnet is then moved out of position so that the pole gap is essentially infinite, and the true remanence, M_r, is measured.*

* The ability to make this kind of measurement is a virtually decisive

In a typical case the data found at a gap of 2.74 cm, and given in centimeters of deflection on the ballistic galvanometer are as follows:

$$\frac{M_r}{M_r'} = \frac{2.66}{2.76} = 0.964$$

We may then find corrected values of M from the observed values of M_{app} by multiplying by 0.964. A different correction must, of course, be used if the pole gap is changed. The data given here have all been corrected for the image effect.

5. Calibration

The chief objective of this work is to measure the change of magnetization produced by a molecule of adsorbate. This does not require anything more than a ratio of magnetizations before and after adsorption. But for the interpretation of these data in terms of electronic interaction we must know the absolute magnetizations. In principle these may be found from a consideration of the geometry and constants of the experiment, but a much simpler procedure is to use as calibrating agent a sample the magnetization of which is precisely known. Pure nickel is suitable for this purpose.

A sample of powdered, polycrystalline nickel* is mixed with silica gel to prepare a pellet similar in size and volume concentration of nickel to those used for the adsorption studies. The sample is heated, *in situ*, in hydrogen for 12 hours at 350°. While this step may be thought to be scarcely necessary, it may result in a small but definite increase of magnetization, suggesting some superficial oxidation in the sample as obtained.

At high fields the approach to saturation of a ferromagnetic is described, as shown by Akulov[9] and by Gans,[10] by a $1/H^2$ law, as follows:

$$M = M_s(1 - b/H^2) \qquad (4.5)$$

where b is a constant. This was found to hold in the present case. One need, therefore, only measure the galvanometer deflections for

---

argument in favor of the Weiss-Forrer experimental method. However, there is some possibility that an error can arise from the change of permeability with H.

* Obtained from the International Nickel Company, New York, New York.

a given mass of nickel at several fields, extrapolate to $1/H = 0$,* and equate the deflection so found to the known value of M_s at the temperature of calibration.

6. General Procedure

The chief requirements for the saturation studies are that the ferromagnetic particles should be small enough so that an appreciable quantity of adsorbate may be taken up, yet not so small that difficulty is encountered in extrapolating the measured magnetizations to obtain a reasonably accurate value of M_0. It is not a requirement for the success of saturation studies that the sample should exhibit collective paramagnetism. In view of this, preparation procedures will vary widely, depending on the particular adsorbent under study, and the problem at hand. Commercially available nickel-kieselguhr catalysts containing 30–50% of nickel are generally suitable. Supported cobalt may be made by impregnation of high area silica gel with cobaltous nitrate solution, followed by drying, careful ignition, and reduction. Experienced workers in the field of heterogeneous catalysis will be able to think of various preparative procedures to try.

Analysis for the total quantity of metal present, either as metal or combined, will offer no difficulty, but accurate determination of the fraction of reduced metal may prove troublesome. In the case of nickel there are several analytical procedures which may be tried, and compared. One method is to measure the volume of hydrogen taken up during reduction. This may be done by circulating a measured volume of hydrogen over the weighed sample in a closed system.[11] The water formed is frozen in a trap. This procedure is reasonably satisfactory except that the catalyst support may contain an appreciable amount of residual water which is slowly released at the temperature of reduction. It must also be remembered that some hydrogen will be chemisorbed on the metal as it is formed. For very highly dispersed nickel this may amount to 20% of the whole volume of hydrogen used.

Another method often used for nickel is to place the weighed sample, after reduction, in hydrochloric acid. The hydrogen displaced is collected and measured.

A third chemical method,[12] applicable to nickel is based on the

* Or to $1/H^2 = 0$. See C. P. Bean and I. S. Jacobs, *J. Appl. Phys.* **31**, 1228 (1960).

von Wartenburg reaction of sulfur vapor with nickel oxide to form sulfur dioxide and nickel sulfide; with nickel metal the reaction is simply the formation of the sulfide. The sulfur dioxide may be determined iodimetrically, or it may be oxidized in hydrogen peroxide to sulfuric acid, which is then titrated. This method appears to be the most nearly reliable for our purpose, although Eggertsen and Roberts[12] express little confidence in the method as applied to nickel supported on alumina. Some difficulty may be experienced because at the temperature of reaction, namely, 850°C, some residual water from a silica catalyst support may react with the sulfur to form sulfur dioxide and hydrogen sulfide. This may be detected by the formation of colloidal sulfur in the effluent. A method for combatting this difficulty is to preheat the sample to 650°C in an inert atmosphere, but this may cause some changes in the proportion of metal present.

Under certain circumstances a determination of the saturation magnetization, M_s, may be expected to yield a satisfactory estimate of the fraction of reduced metal. Knowing the total mass of metal present both in reduced and oxidized forms we may readily calculate what M_s should be if the reduction is 100% complete. If M_s (obs) can be found, then the fraction, f, of reduced metal present is simply M_s (obs)$/M_s$. This procedure has been used by Heukelom et al.[13] and by Trzebiatowski and Romanowski.[2] These two studies have considerable bearing on our overall problem. They will be described in some detail.

Heukelom et al.[13] determined the degree of reduction by the sulfur vapor method as applied to nickel-silica samples prepared by van Eijk van Voorthuysen and Franzen.[14] They then measured the magnetization over a range of field strength at room temperature, and obtained the saturation magnetization by the extrapolation method previously described. The degree of reduction, in samples reduced at various temperatures and for various times, was then compared as found by the chemical versus the magnetic method. The results so obtained were in reasonably satisfactory agreement, the maximum difference appearing to be about ±10% of the average of the two methods.

Trzebiatowski and Romanowski[2] made a similar study for nickel supported on high-area alumina, but they used the hydrochloric acid method to compare with the magnetic. The agreement here was also reasonably satisfactory, the maximum deviation occasionally rising to ±20% of the mean, but being within ±3% for most of the samples except those containing quite low over-all concentrations of nickel. Similar studies on both supported and unsupported nickel have been made by Umemura.[15]

The magnetic method for obtaining the fraction of reduced metal, as described above, depends on two assumptions. One is that the extrapolation from data obtained at room temperature will give a reasonably accurate measure of M_s. The other assumption is that M_s for very small particles is the same as that for massive metal. Concerning the second assumption we shall have more to say later. But the validity of the long extrapolation to obtain M_s is now subject to experimental test. Dietz[8] gives data on a moderately sintered sample of commercial nickel-kieselguhr containing about 52% of nickel. This shows a normal saturation magnetization, M_0, obtained from measurements over a range of field at 4.2°K. The precision is within ±1.0%. If $\bar{\mu}_{Ni}$ for small particles is the same as that for massive nickel we must regard this sample as being essentially completely reduced to metal.

It so happens that this same sample showed no remanence at 300°K. The theory of collective paramagnetism may, therefore, be applied; at least this is true if the measurements are made at or above this temperature. The procedure is then to take the magnetizations, M, as a function of field strength and, as done by Trzebiatowski, use the empirical relation given by Heukelom [Eq. (3.16)] but in the form*

$$\frac{1}{M} = \frac{1}{M_s} + \frac{1}{M_s(\alpha H)^{0.9}} \tag{4.6}$$

If $1/M$ is plotted against $1/H^{0.9}$ it is possible to obtain a value for M_s by a straight line extrapolation to $1/H^{0.9} = 0$. This extrapolation is shown in Fig. 25. In this way it is found that M_s as found by the Heukelom-Trzebiatowski method is about 9% lower than that of pure massive nickel at the same temperature. This agreement is reasonably satisfactory. But although the data have, in this case, been corrected for both demagnetization and image effects, the precision is not sufficient for our principal purpose, which is to measure the effect of adsorbed molecules. These effects rarely

* Our M and M_s are the same as Heukelom's σ and σ_∞ multiplied by the density.

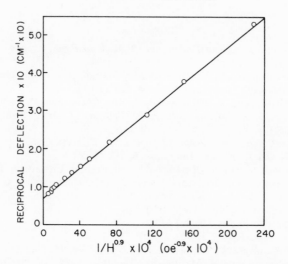

FIG. 25. A plot of M vs. $1/H^{0.9}$, using the data obtained by Dietz[8] and his value of the true saturation magnetization at $0°K$, namely, M_0. This procedure permits a comparison of M_0 obtained by the Heukelom[13]-Trzebiatow-ski[2] extrapolation with that actually measured at high fields and low temperatures.

exceed 20% for complete surface coverage. It appears, therefore, that there is no real substitute for a determination of M_0 based on data obtained at high fields and at $4.2°K$, or lower.

This section will be concluded with some remarks concerning precision. The absolute saturation magnetizations approach a precision of $\pm1\%$, or better under favorable conditions. But the relative magnetizations before and after vapor adsorption are accurate to about $\pm0.1\%$. A typical nickel-silica sample is capable of chemisorbing nearly 20 cc of hydrogen per gram of nickel at room temperature. Complete coverage cannot be utilized because of the necessity for keeping the quantity of gas in the dead space negligible. The dead space, including a turntable McLeod gauge was 185 cc. As closed off during actual measurement the dead space was about 92 cc. The volume (STP) of hydrogen taken up in a typical experiment was of the order of 8 cc per gram of nickel, and this could be measured with at least the same precision as could the magnetization. Over-all precision in determining the change of magnetization per cubic centimeter of gas adsorbed is thus rather better than $\pm1\%$ in favorable systems, less in others.

REFERENCES

1. J. J. Broeder, L. L. van Reijen, and A. R. Korswagen, *J. chim. phys.* **54**, 37 (1957).
2. W. Trzebiatowski and W. Romanowski, *Roczniki Chem.* **31**, 1123 (1957).
3. W. Romanowski, *Roczniki Chem.* **34**, 239 (1960).
4. J. A. Sabatka and P. W. Selwood, *J. Am. Chem. Soc.* **77**, 5799 (1955).
5. E. L. Lee, J. A. Sabatka, and P. W. Selwood, *J. Am. Chem. Soc.* **79**, 5391 (1957).
6. C. P. Bean and I. S. Jacobs, *J. Appl. Phys.* **27**, 1448 (1956).
7. P. Weiss and R. Forrer, *Ann. phys.* **5**, 153 (1926).
8. R. E. Dietz, Doctoral Dissertation, Northwestern University, Evanston, Illinois, 1960. Available from University Microfilms, Inc., Ann Arbor, Michigan. Also, M. B. Stout, "Basic Electrical Measurements," pp. 370–373. Prentice-Hall, New York, 1950.
9. N. Akulov, *Z. Physik* **69**, 822 (1931).
10. R. Gans, *Ann. Physik* **15**, 28 (1932).
11. F. N. Hill and P. W. Selwood, *J. Am. Chem. Soc.* **71**, 2522 (1949).
12. F. T. Eggertsen and R. M. Roberts, *Anal. Chem.* **22**, 924 (1950).
13. W. Heukelom, J. J. Broeder, and L. L. van Reijen, *J. chim. phys.* **54**, 474 (1957).
14. J. J. B. van Eijk van Voorthuysen and P. Franzen, *Rec. trav. chim.* **70**, 793 (1951).
15. K. Umemura, *Nippon Kagaku Zasshi* **81**, 863, 866, 991 (1960).

CHAPTER V

Saturation Data

1. Some Experimental Results

In this section there will be presented the data thus far obtained on the change of atomic magnetic moment, $\bar{\mu}_A$, in the adsorbent as hydrogen is taken up. First, the purpose of the work will be stated in more explicit terms than heretofore. Later sections will be devoted to examining the validity of the procedure, and to a discussion of the kinds of information obtained from similar measurements on related systems. But such conclusions concerning bond type as may be drawn will be deferred until after presentation of the low-field data.

The measurements of magnetization at low temperature and high field yield, in favorable cases, reasonably accurate data for the saturation magnetization at absolute zero, M_0. They also give the change in saturation magnetization, ΔM_0, which occurs when vapor molecules are admitted to the surface of the adsorbent.* Let N_A be the number of metal atoms, of average moment $\bar{\mu}_A\beta$, per unit volume of sample. Then

$$M_0 = N_A\bar{\mu}_A\beta \tag{5.1}$$

and

$$\Delta M_0 = \Delta(N_A\bar{\mu}_A\beta) \tag{5.2}$$

whence the fractional change of magnetization which is produced during chemisorption is:

$$\frac{\Delta M_0}{M_0} = \frac{\Delta(N_A\bar{\mu}_A)}{N_A\bar{\mu}_A} \tag{5.3}$$

---

* Perhaps it should be stated again that while the measurements of saturation magnetization are made at liquid helium temperatures, the adsorbate is admitted at room temperature or higher. Virtually no vapor remains in the gas phase during actual measurements.

We shall not beg the question as to whether chemisorption causes a change of N_A or of $\bar{\mu}_A$; either or both are possible.

Let ϵ be the apparent change in $\bar{\mu}_A$ caused by one atom of hydrogen adsorbed per atom of adsorbent. Then on a sample of volume V containing VN_A atoms of adsorbent the total change of magnetization would be:

$$VN_H\epsilon = \Delta(N_A\bar{\mu}_A)V \tag{5.4}$$

where N_H is the number of atoms of hydrogen taken up by a sample of unit volume. Whence, from Eq. (5.3):

$$\epsilon = \frac{(\Delta M_0/M_0)N_A\bar{\mu}_A}{N_H} \tag{5.5}$$

It will be noted that ϵ has the units of the Bohr magneton. The significance of this important quantity will be discussed below.

A preliminary attempt[1] at direct measurement of the quantity ϵ for nickel gave a value of 0.8 Bohr magneton but, in view of the low precision, this gives little more than an order of magnitude. Figure 26 shows saturation data obtained by Dietz[2,3] on a sample

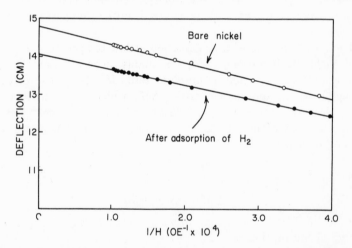

FIG. 26. The approach to saturation magnetization for a nickel-kieselguhr sample at 4.2°K before and after the adsorption (at 25°C) of 7.76 cc (STP) hydrogen per gram of nickel (after Dietz).

of commercial nickel-kieselguhr catalyst manufactured by Universal Oil Products Company. This had been reduced at 350°C for 12 hours, heated for several hours in helium at 600°C to cause a moderate increase of particle size, then cooled to the temperature of measurement which was 4.2°K. The weight of nickel in the sample was 0.1534 g. Data are also shown for the same sample after it had adsorbed 1.19 cc (STP) of hydrogen. The saturation magnetization of this sample prior to admission of the hydrogen was 97% of that of massive nickel. The mean particle radius derived from low-field data $(\bar{v}^2/\bar{v})$ was 64 A.

From Fig. 26 the corrected galvanometer deflections (which vary directly with magnetization) show that $\Delta M_0/M_0 = 0.72/14.78 = 0.0487$. We shall assume that $\bar{\mu}_{Ni} = 0.606$, although this matter will be discussed further below. Then ϵ may be calculated as follows:

$$\epsilon = \frac{4.87 \times 10^{-2} \times 0.606\ \beta \times 0.1534\ \text{g} \times 22.4 \times 10^3\ \text{cm}^3\ \text{mole}^{-1}}{1.19\ \text{cm}^3 \times 2 \times 58.71\ \text{g mole}^{-1}}$$

$$= 0.72\ \beta \tag{5.6}$$

The average of several determinations on similar samples with moderately varying quantities of adsorbed hydrogen gave an average ϵ of 0.71. The average M_0 for these samples was 98.5% of that for massive nickel. The choice of moderately sintered samples for study was dictated by the convenient particle size which was small enough to adsorb a fairly large amount of hydrogen yet large enough to permit an accurate extrapolation to $1/H = 0$, to obtain M_0.

Later results on nickel-alumina samples containing smaller particles of radius 25 A from $\bar{v}$, have given values of ϵ down to about 0.35β.* Similarly, measurements on cobalt particles by C. R. Abeledo in the writer's laboratory give $\epsilon = 0.5\beta$ for the cobalt-hydrogen system.

Sufficient saturation data are not available to permit a definite statement concerning the possible dependence of ϵ with changing surface coverage. There is, likewise, as yet no definite information concerning the possible variation of ϵ with particle size. We shall, however, have something more to say concerning these matters in a later chapter.

One additional set of data will be useful later. This is the plot of M against H/T for a coprecipitated nickel-silica sample before

---

* Data obtained by D. Reinen.

and after adsorption of hydrogen. These data are shown in Fig. 27. It will be noted that superposition over the range 77°–300°K is just as satisfactory for the sample *after* adsorption as before; the whole curve merely being displaced to a lower M as expected.

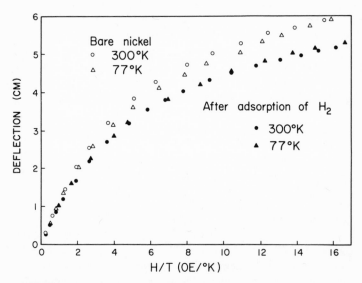

Fɪɢ. 27. Plot of galvanometer deflection (proportional to M) against H/T for a nickel-silica sample before and after adsorption of hydrogen at 25°C (after Dietz).

These data were obtained in the sequence 300°, 77°, and 300°. There was no evidence of hysteresis between first and last measurements.

In all the measurements reported there was no change of magnetization, at a given temperature, over a period of hours and, in a few cases investigated, over a period of days.

2. Discussion of the Experiment

Before we draw any conclusions about the quantity ϵ and its significance in surface chemistry it will be necessary to examine in more detail the several assumptions implicit in the experiment. The first of these assumptions is that $\bar{\mu}_A$ is a constant over the range of conditions encountered. This is to say that the metal particles are essentially free from impurities (such as copper in nickel) which would alter the moment; that $\bar{\mu}_A$ is independent of particle size; and that the silica support has a negligible effect on $\bar{\mu}_A$. These several problems will be discussed in turn.

The possible presence of dissolved impurities of such a nature that they might change $\bar{\mu}_A$ is readily dismissed. Chemical and spectroscopic analysis shows that there is no element present in a quantity sufficient to cause a measurable change in magnetization.

The question of the influence of particle size on $\bar{\mu}_A$ is more serious. It is obvious that at some state of subdivision there will be a breakdown of the exchange interaction which leads to spontaneous magnetization. A single atom cannot exhibit collective paramagnetism. There seems to be no evidence that the saturation moment per atom of adsorbent is any different (for cobalt, at least) down to particles only a few ångstroms in diameter. The question is discussed by Bean and Livingston,[4] who quote data obtained in the General Electric Research Laboratories on particles having about 25% of all their atoms on the surface. Similarly, no evidence of such an effect has yet been found in the writer's laboratory. The General Electric experiments do, however, differ in one respect from our own. The cobalt and iron particles were in a matrix of copper and of mercury, respectively. If surface states exist, and if they bring about some change in the average magnetic moment of the particle, it would seem that the effect might be rather different when the interface is metal-metal as opposed to metal-vacuum or metal-silica. In the meantime we shall assume that $\bar{\mu}_A$ is independent of particle size.

There are at least two other ways in which the particle size may have a complicating effect on the saturation results. One of these is the possibility that the Curie temperature, T_c, may be a function of particle size. Reference to this problem was made earlier and it will now be examined in more detail, although it affects the nonsaturation data (to be described later) more than it does the work under discussion. At an early stage in these investigations it was thought that the progressive fall in average coordination number would cause a change in T_c and that this would become important as soon as a few per cent of all atoms in a particle had positions on the surface. There is also the possibility that T_c would suffer a further change if molecules of a vapor were adsorbed on the surface. If effects such as these occur they might interfere with interpretations given to the saturation data, provided that lowering of the Curie temperature made a significant change in the spontaneous magnetization at temperatures as low as the liquid helium region.

This question is also discussed by Bean and Livingston.[4] Some of the evidence tending to establish the existence of such an effect has been obtained on thin films, but Bean has shown that the results on thin films are not applicable to particles such as those under discussion. There has, thus far, been no development of a theory of spontaneous magnetization in very small ferromagnetic particles. Particles of cobalt down to about 14 Å diameter have been shown experimentally by Bean et al.,[5] Cahn et al.,[6] and Knappwost and Illenberger[7] to have normal magnetization curves. These particles were obtained as precipitates of cobalt from dilute cobalt-copper solid solutions. The only reservation one might have in connection with these results is that the maximum temperature reached in the measurements was only a small fraction of the absolute Curie temperature of cobalt, namely, 1400°K. On the other hand, Henning and Vogt[8] have reported subnormal Curie temperatures for small particles of iron, and Kneller[9] has found similar results for 27 Å particles of Ni_3Mn.

As pointed out by Bean and Livingston,[4] the problem of T_c determination in a specimen exhibiting collective paramagnetism may be quite difficult. A reason for this is that the field necessary to produce a significant orientation of the particles may be sufficient to modify the moment of the particle. A possible solution to the problem is, as suggested by the same authors, to rely on Eq. (3.10) (p. 45) wherein the initial susceptibility (at low field) is seen to be proportional to the square of I_{sp}. But in a sample showing a wide distribution of particle diameters Eq. (3.10) tends to minimize the effect of smaller particles which do not contribute very much to the magnetization as obtained at low fields.

A more satisfactory procedure[10] is to consider the complete expression:

$$\frac{MM_0}{I_{sp}} = M_0 V \int_0^\infty \left[\coth\left(\frac{vM_0}{k} \cdot \frac{I_{sp}H}{M_0T}\right) - \frac{kTM_0}{vM_0I_{sp}H} \right] f(v)dv \quad (5.7)$$

One may then plot MM_0/I_{sp} vs. $I_{sp}H/M_0T$, using known values of I_{sp}/M_0. This is shown in Fig. 28 for a nickel-silica sample prepared by impregnation of silica gel with nickel nitrate solution, followed by drying and reduction in hydrogen. The particle diam-

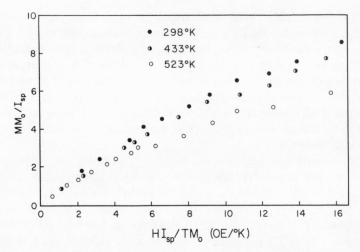

FIG. 28. Magnetization plotted against H/T for nickel-silica. I_{sp} has been taken as that for a ferromagnetic substance having the normal Curie point of nickel (after Abeledo).

eter derived from $\bar{v}$ was 30 A. It will be noted that the measurements were carried out up to 250°C, only about 100° below the normal Curie temperature. Superposition of the data is not found

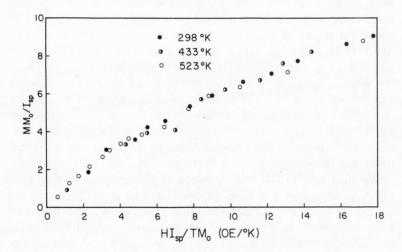

FIG. 29. The data of Fig. 28 replotted for a ferromagnetic substance with Curie point 70° lower than that (635°K) of normal nickel (after Abeledo).

but, if the I_{sp}/M_0 ratios used are those of a ferromagnetic with Curie temperature $565°K$, instead of the normal $635°K$ (see Fig. 29), then satisfactory superposition occurs. Similar data with other preparations having diameters up to 85 A showed a similar small lowering of the Curie point. It should be emphasized that no lowering of M_0 was observed in these samples. This seems to rule out the effect of dissolved impurities. The Curie point lowering seems not to be a function of particle size for any of the several preparations investigated.

Our conclusions with respect to the T_c problem are, therefore, that no very large effect occurs down to 30 A diameter particles of nickel. The effect is not a function of particle size, and it has only a minor influence on our conclusions concerning the significance of the quantity ϵ, and on the interpretations of low-field data to be presented later. Nevertheless, the effect is peculiar, and is one which might merit further investigation.

The results given above, namely, that ϵ is only slightly larger than $\bar{\mu}_{Ni}$, suggest that a particle of nickel small enough to have all, or nearly all, its atoms on the surface would become nonmagnetic on the formation of a complete monolayer of adsorbed hydrogen. If this phenomenon occurs to any extent it could seriously affect the apparent value of ϵ. A somewhat parallel effect may actually occur for the palladium-hydrogen system in which no further decrease of magnetic susceptibility (as measured at room temperature) occurs if the atomic ratio Pd:H falls below 1:0.6. If an effect like this occurs for nickel we would expect that ϵ would show some diminution with increasing surface coverage. No such diminution has been observed for hydrogen sorptions running up to 19 cc (STP) of hydrogen per gram of nickel in a coprecipitated sample for which $\bar{v}$ corresponded to a particle radius of only 10 A. Unfortunately, these data do not quite settle the matter because the extrapolations to $1/H = 0$ are somewhat uncertain for such very small particles.

We turn now to the final question in this group, namely, what effect, if any, does the silica support have on $\bar{\mu}_A$? This is related to the problem of "surface cleanliness" which will be discussed first. Surface cleanliness is one of the most persistently debated questions in surface chemistry. On the one hand, it is pointed out that little quantitative information may be obtained relative to a surface the composition of which is unknown. On the other hand, it is obvious that solid surfaces used in actual catalytic practice are far from being uncontaminated. There is some evidence that the metal surfaces used in the investigations under discussion are relatively free from surface contamination. This statement may seem to be a rather surprising one. It will be supported in some detail.

Dietz[2,3] has shown that a pellet of nickel-silica evacuated to a minimum pressure of 10^{-6} mm Hg may be no more contaminated than a nickel film at 10^{-10} mm. The reason for this is that the surface area of nickel metal in a typical nickel-silica pellet weighing, say, 1 g may be at least 10 m², while that of a typical film used in adsorption studies may be at least 10^3 times smaller. Furthermore, this difference in surface area may be accentuated by the nature of the diffusion process necessary for a gas molecule to move inside the pellet. Gas molecules are able to reach the metal in the pellet only by passing the geometric boundary of the pellet and this may be a surface of no more than 1 cm³. The number of molecules entering the pellet per unit time may thus be less than the number striking the surface of a typical film. Yet once inside the pellet the vapor molecules find themselves faced with a metal surface many orders of magnitude larger than that of the film. To state this in another way, the metal-silica pellet acts as a "getter" for any gas molecule which has crossed its geometric boundary.

A rather different way to present the same argument is as follows: let us say that the sample chamber has a dead-space of 100 cc and that this is held at a pressure of 10^{-6} mm Hg. The total quantity of contaminating gas present is then far too small to make any measurable effect on a gram or two of nickel possessing a specific surface in the neighborhood of 10 m² or more. But the same quantity of gas could have a very serious effect on a film of which the surface was only 100 cm².

The above argument does not apply to contaminants which may not have been removed in the preparation stage or which may emerge from the silica support in the same manner as water may emerge from the glass on which a metal film is condensed. There are two lines of evidence which tend to show that neither of these sources of contamination is serious. The first is that monolayer coverage of nickel lowers the saturation magnetization by about 10% in a typical nickel-silica sample. The fact that $\bar{\mu}_{Ni}$ for the nickel in these samples is the same, within ±1%, as that of pure, massive nickel suggests that no more than one-tenth of the surface could be contaminated.

There is quite another line of evidence which tends to confirm the views expressed above. It is well known that hydrogen is chemisorbed virtually instantaneously on nickel surfaces in the temperature range from about −150°C to several hundred degrees above room temperature. This is followed by a slower hydrogen sorption, the nature of which is still somewhat obscure, but concerning which we shall have more to say later. Schuit and de Boer[11] have made the point that the slow effect, which is negligible on metal films, is also negligible on exhaustively reduced nickel-silica systems. There is no essential difference in the kinetics of adsorption on a film and on a supported metal, provided that the latter has been prepared with appropriate care. The inference is that if the film is free from contamination the supported metal is likewise free. This view receives confirmation from the experiment in which Schuit and de Boer deliberately contaminate the surface of the supported metal with oxygen, and show that this contamination causes a marked increase in the volume of "slow" hydrogen taken up, at the expense of the "instantaneous" hydrogen.*

These several lines of evidence argue for the surface cleanliness of supported metal systems provided, of course, they have not been allowed to stand for any appreciable length of time after reduction, and especially have not been heated in vacuum or in inert atmosphere longer than is necessary to remove the residual adsorbed hydrogen after the reduction step. By contrast, the evidence that films, as ordinarily prepared, are similarly free from contamination is all circumstantial evidence. Films are sometimes handled in air after "protection" of the surface by evaporated silicon monoxide. The writer believes that this procedure is indefensible as actually consisting of gross and complete contamination of the surface. Silica gel, used as a metal support, always contains

---

* Low field, permeameter studies by the author, as described in the following chapter, have shown that prolonged evacuation at elevated temperatures leads to a marked decrease in the quantity ε. This is comprehensible if we may consider that in the absence of an appreciable hydrogen pressure it is possible for residual water in the silica support to react with the nickel surface to form a superficial oxide layer. This reaction does not, however, normally occur at room temperature because of the extraordinarily low equilibrium pressure of water vapor over the silica gel under these conditions.

a trace of water. It might be thought that this water would emerge from the silica and contaminate the metal either as adsorbed water molecules or as oxide and hydrogen. This process probably occurs at elevated temperatures, but we shall later present some virtually conclusive evidence that progressive contamination does not readily occur at room temperature in a good vacuum system.

The possibility that the silica support may exert some influence on the electron distribution in small supported metal particles is one which cannot be dismissed lightly. This effect, if it exists, might be considered a kind of contamination by the support. Films should similarly show such an effect unless they are prepared, as rarely occurs, free from any glass or other surface on which the film is condensed. There is a fair amount of evidence dealing with this problem; we shall be concerned with possible influences of the support on $\bar{\mu}_A$. This presupposes that the metal particles are actually attached somehow to the support. There is a possibility that the particles are executing a Brownian movement in some sort of silica prison cell.

Interest in this area comes from the often-repeated observation that the catalytic activity of a supported metal depends, at least in part, on the nature of the support. Evidence has also been presented by Eischens and Pliskin[12] to show that the infrared absorption spectrum of chemisorbed carbon monoxide on supported platinum is displaced by a change of catalyst support from silica to alumina. On the other hand, the bands attributed to the Pt-H group appear to be independent of the nature of the support[13]; and studies of the K X-ray absorption edge by Lewis[14] fail to show that alumina-supported nickel is in any way different from massive nickel.

The study most closely related to the problem at hand is that of Schwab et al.[15] This paper has been quoted as giving proof of the effect of the support on the electron band structure of supported nickel. The work will, therefore, be examined in some detail.

The procedure used by Schwab et al. was to start with high area alumina, samples of which were "doped" with TiO_2, GeO_2, BeO, and NiO. These samples were prepared by mechanical mixture followed by sintering at 1050°C for 4½ hours. Measurements of electrical conductivity at 900°C were made on the doped supports before addition of the nickel metal. The nickel was then added by condensation from evaporated nickel. Analysis was done by dissolving the sample in nitric acid, heating with hydrochloric acid, then treating with bromine-water, ammonia, ethanol, and dimethyl-glyoxime for colorimetric determination of the nickel. No statement is made as to how the nickel metal was differentiated from the nickel oxide in the sample doped with NiO. The proportions of nickel so found were all about 0.1% as shown in Table VI.

TABLE VI

EFFECT OF DOPED AL$_2$O$_3$ SUPPORTS ON THE MAGNETIC MOMENT OF NICKEL
AS REPORTED BY SCHWAB, BLOCK, AND SCHULTZE

Sample	Electrical conductivity of support at 900°, $\times 10^6$ (ohm^{-1})	Weight of Ni $\times 10^5$ (g)	μ_{Ni}
Ni on Al$_2$O$_3$ + 5 mol % TiO$_2$	7.1	4.0	0.46 ±0.004
Ni on Al$_2$O$_3$ (pure)	4.2	1.3	0.51 ±0.005
Ni on Al$_2$O$_3$ + 5 mol % NiO	2.5	4.0	0.59 ±0.006

Our concern is with the magnetic data and the reported influence of the support on μ_{Ni}, but attention should be drawn to the astonishingly small change of electrical conductivity for such large proportions of doping agent. By contrast, the conductivity of nickel oxide may be changed many orders of magnitude by a trace of lithia.

If the results shown in Table VI are valid they establish a very interesting phenomenon. Further information concerning experimental procedure and method of interpretation is given in the Dissertation of Schultze[16] to which the reader is referred for details in ref. 15 quoted above. The magnetic measurements were made by the Faraday method, and what precautions were taken to prevent access of air to the samples during transfer and measurement are not stated. The measurements were made at several fields to a maximum of 4540 oersteds, and at 77°, 195°, and 293°K.

The data are shown (Fig. 9 of ref. 16) as susceptibilities plotted against reciprocal field. These data lie on curves, and a long extrapolation is then made to $1/H = 0$, apparently in the expectation of obtaining the saturation magnetization. It is, of course, true that the magnetization of a ferromagnetic specimen approaches the saturation magnetization as the reciprocal field approaches zero. But this is certainly not true of the *susceptibility* of a ferro-magnetic which must obviously approach zero at infinite field. A reasonable procedure would have been an attempt to extrapolate not the susceptibility but rather the magnetization to infinite field as was done by Broeder *et al.* and by Trzebiatowski and Roman-owski as described on p. 47. But, as pointed out on p. 64 even this procedure could scarcely yield $\bar{\mu}_{Ni}$ values sufficiently precise for the purposes under discussion. We are, therefore, forced to the conclusion that no effect of the catalyst support on the saturation moment of nickel (or of any other metal) has yet been established.

A more fruitful line of investigation, and one which has some bearing on our overall problem has been carried out at the labora-tories of the General Electric Company. This work is described by Bean and Livingston.[4] These experiments, to which reference has already been made several times, consist of magnetization meas-urements on precipitated cobalt from solid solution in copper, and also work (by Luborsky) on small particles of iron suspended in mercury. The particle radii range from 21 to 77 A. For the smaller particles about 25% of the cobalt atoms must be on the surface of the particle and are, presumably in effective contact with copper atoms. The saturation magnetization of these small particles of cobalt is the same, within 2%, throughout the diameter range in spite of the fact that in the larger particles the fraction of cobalt atoms in contact with copper atoms must be much less than 25%. Similar particles covered with a monolayer of chemisorbed hy-drogen would suffer at least a 10% decrease of saturation mag-netization.

As Bean points out, there is no real discrepancy here. A mono-layer of vapor molecules on a metal surface must be subject to extraordinarily high polarization. A monolayer of copper atoms on a cobalt surface would, presumably, suffer the same effect. But the experiment described above is quite different. Those copper atoms touching cobalt atoms are also touching copper atoms as part of a continuous lattice. The only tendency to cause redistribution of electrons at such an interface would be the difference in work function of the two metals. Bean[17] has shown that this would have a negligible effect on the saturation magnetization of the cobalt; and, presumably, of the iron precipitate particles suspended in mercury.

If copper and mercury in contact with small ferromagnetic particles fail to produce a measurable change of $\bar{\mu}_A$, then it is difficult to see how silica and alumina could do so, whether rendered semi-conductive or not. Our conclusion with respect to this point is, therefore, that no such electronic redistribution has yet been proved. An explanation which occurs to the writer for the observations of Eischens and others on the infrared absorption spectra of adsorbed molecules, and for the apparent effect of the support on catalytic activity, is that the small particles of metal are highly strained and that the anisotropy so produced is reflected in the properties mentioned. This effect, if present, would not affect the saturation magnetization but it might very well affect the permeability and the remanence. In this connection there should be mentioned the experiment of Neugebauer[18] who found that nickel films, prepared under vacuum conditions less than 10^{-9} mm Hg, showed a loss of saturation magnetization on exposure to hydrogen provided that the film thickness was not over about 30 A. Somewhat thicker films showed a loss of magnetization on exposure to hydrogen or on being covered with a condensed film of copper; but on these thicker films the whole effect could be attributed to a change in anisotropy energy. These observations by Neugebauer suggest that effects other than electron spin pairing may be important in certain aspects of chemisorption.

Two other points remain to be mentioned. The first is the well-known slow sorption of hydrogen which takes place after the virtually instantaneous take-up is complete. We shall have something to say concerning this process later. For the saturation measurements thus far described the slow process appears to be of minor significance. The reason for this is that the surface coverage is not carried beyond that corresponding to about 0.1 mm Hg pressure. There is at no time any appreciable quantity of hydrogen present and capable of being taken up by the slow process.

The other point is this: can we be sure that information gained about the metal-hydrogen bond at 4°K bears any relation to the bond type at, say, room temperature? No complete answer can yet be given to this question, but we can state unequivocally that no appreciable change of bond type (as reflected in a change of I_{sp}) occurs in the range 77°–300°K. This is shown in Fig. 27 taken from the data of Dietz.[2] Superposition of experimental points in a plot of M vs. H/T could occur for different temperatures only if the two or more proposed types of bonding had the same electron spin-pairing effect. This possibility seems unlikely. This experiment does not exclude the possibility that two or more kinds of bonding occur. It shows that any change in relative proportions of bond types is improbable.

3. The Significance of ϵ

In this section there will be discussed the meaning of the quantity ϵ, and its possible relation to information available on other systems which appear to be related to the metal-hydrogen adsorbent-adsorbate system.

The only systems for which saturation data are available at this writing are the nickel-hydrogen system and the cobalt-hydrogen system. For these systems the quantity ϵ, the change of atomic magnetic moment per atom of hydrogen adsorbed, is about -0.7 Bohr magneton for the nickel, and -0.5 for the cobalt, for the measurements in which most confidence may be placed. The change may be somewhat smaller for samples containing smaller particles.

If the magnetic moment of nickel is due to electron spins only, then $\bar{\mu}_{Ni}$ is numerically equal to the number of unpaired electrons and ϵ might be referred to as the number of electron spins paired per atom of hydrogen adsorbed. But Argyres and Kittel[19] have shown that the number of effective electron spins contributing to the saturation magnetization is $\bar{\mu}_{Ni} (2/g)$, where g, as before, is the splitting factor. The splitting factor for colloidal nickel has been measured by Bagguley,[20] and confirmed by others, as $g = 2.2$. The saturation moment of nickel appears to be 0.606β, whence[21] the number of unpaired electron spins is 0.54, rather than the formerly accepted value of 0.61.

If the change of magnetization caused by adsorbed hydrogen arises solely from pairing of electron spins, then ϵ is numerically the number of spins so paired. But if some change of spin-orbital coupling is involved, such as recently suggested by Dowden,[22] then the g factor may be expected to change and the number of electron spins paired per atom of hydrogen adsorbed may be somewhat less than ϵ. Hollis and Selwood[23] have reported ferromagnetic resonance measurements on impregnated nickel-silica preparations comparable to those described above. The g factor is, indeed, approximately 2.2 (the resonance line is quite broad) and the g factor does not change when hydrogen is adsorbed on the nickel, although the integrated line intensity is diminished as expected, as shown in Fig. 30.

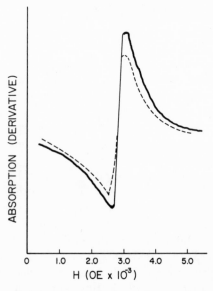

FIG. 30. Ferromagnetic resonance absorption of a 10% nickel-silica preparation before (—) and after (---) adsorption of enough hydrogen to approach saturation of the nickel surface.

Further discussion of these results will be presented below; but reference will first be made to the interpretation of magnetic data given by Broeder et al.[24] These authors measured the magnetization of nickel-silica preparations as a function of surface coverage with hydrogen. The measurements were made at room temperature, at various fields up to 2000 oersteds. It is claimed that the relative change of magnetization ($\Delta M/M$ in our terminology) is proportional to $\Delta M_0/M_0$ but, as will be described in some detail later, the analysis by Dietz[2,3] shows that this assumption is true only if the sample exhibits true collective paramagnetism, and also if the fractional surface coverage is moderate. Inasmuch as both of these requirements may not have been met in the experiments of Broeder et al. we may proceed to further examination of their conclusions with some reservations.

These conclusions are presented in the following way: let γ be the fraction of nickel atoms on the surface of a particle, and let it be assumed, arbitrarily, that each surface atom may adsorb one hydrogen atom with consequent loss of 0.6 Bohr magneton by the particle. Hence $\Delta M/M = -\gamma/0.6 = -1.67\gamma$. The actual $\Delta M/M$ observed for sorption of 50.6 cc (STP) of hydrogen per gram of nickel, corresponding to $\gamma = 0.31$, was 0.22, rather than the expected value of 0.52. This conclusion is interpreted in terms of a substantial covalent character of the Ni-H bond.

While no quarrel with this conclusion is called for, yet it is doubtful if much confidence may be placed in the reasoning by which the conclusion was reached. According to the saturation data given above $\Delta M_0/M_0$ is possibly slightly larger, rather than considerably smaller, than the normal moment of nickel. A further difficulty is the implied assumption that because $\Delta M/M$ is proportional to $\Delta M_0/M_0$ the proportionality constant is unity. It will be shown later that for measurements made at relatively low values of M/M_s, as these were, the proportionality factor is more nearly 2.

Some enlightenment on the nickel-hydrogen bond problem might be expected from a consideration of the magnetic properties of other nickel systems, and of palladium-hydrogen. There are an abundance of data available on nickel-copper alloys, on palladium-hydrogen, and on palladium-silver. The data and some conclusions which have been drawn thereon will be examined next. If the saturation moment, $\bar{\mu}_{Ni}$ of nickel is taken as 0.606 Bohr magneton, and the g factor is 2.20, then the number of unoccupied $3d$ states in nickel is 0.55 per atom. As previously explained (p. 30) this has been interpreted in terms of the band model as meaning that one of the $3d$ subbands is filled and the other as filled except for a 0.54 electron vacancy. The $4s$ band then has 0.54 electron which does not contribute to the magnetic moment. If copper is dissolved in nickel the saturation moment falls linearly, becoming zero at 53 atom % of copper according to Ahern et al.[25] although earlier investigations had given the critical concentration as about 60%.

The copper atom is, of course, diamagnetic and it possesses one more electron than does nickel. It has been thought* that electrons from the copper are added to the unfilled d subband of the nickel in such a way that the band is just filled at 53% copper. According

* A clear presentation of this problem is given by Kittel.[26]

to a recent recalculation by Crangle and Martin,[21] the change produced in $\bar{\mu}_{Ni}$ per atom of dissolved copper is 0.48 Bohr magneton. If this is true then the number of electrons transferred to the unfilled $3d$ subband per solute atom is somewhat less than 0.5.

It will be clear that the above process of substituting a copper atom for a nickel atom to form an alloy is somewhat different from the adsorption of an atom of hydrogen on the surface of nickel. If we expand our definition of ϵ to be the change in $\bar{\mu}_A$ produced by the *addition* of any foreign atom, then ϵ for copper on nickel is 0.46 Bohr magneton, and $\bar{\mu}_{Ni}$ reaches zero at an atom ratio corresponding to NiCu$_{1.17}$. It has generally been thought that the remaining s electrons brought in by the copper are transferred to the $4s$ conduction band.

Before leaving this topic of nickel-copper it will be necessary to refer again to the interesting results obtained by Neugebauer.[18] He found that copper (deposited as a film) on a 67 Å nickel film produced anisotropy changes, but no change of M_s. Nickel particles of diameter about 128 Å (based on $\bar{v}$) gave $\Delta M_s/M_s$ in excess of 10% when covered with what appeared to be a monolayer of hydrogen. This difference *may* be related to some inherent difference between continuous films and isolated particles, but a more obvious explanation is that the available surface area of a 128 Å diameter sphere is several times larger than that of a 67 Å film. It may be noted also that the high-field method for finding the volume represents an upper limit for $\bar{v}$, and therefore a lower limit for the surface area.

In view of the difference between a nickel-copper alloy and a nickel-hydrogen adsorbent-adsorbate system it may be instructive to consider the palladium-hydrogen system. This differs from the nickel-hydrogen system in that the hydrogen actually dissolves in the palladium forming something closely resembling an alloy. Palladium is not ferromagnetic and, so far as is known, never exhibits collective paramagnetism. The paramagnetic susceptibility is fairly large and it becomes zero (at room temperature) at an atom ratio of about PdH$_{0.64}$. Following an idea presented by Mott and Jones,[27] later investigators have assumed that each hydrogen atom donated one electron to the d subband of the palladium. This is based on the belief that the electron distribution in palladium is similar to that in nickel except that we are dealing here with the $4d$ band instead of the $3d$.

There may have been some tendency here to argue in a circle—

assuming that $\mu_{\text{Pd}} = 0.64$ because it takes 0.64 atoms of H to lower the moment to zero, and assuming that each hydrogen donates one electron to the palladium for the same reason. But we are not entirely without other evidence on the moment of palladium, although this is by no means so convincing as that on the saturation moment for nickel. Although palladium is fairly strongly paramagnetic the susceptibility is rather erratic at low temperatures. Wucher[28] has shown that in the temperature range 700°–1600°K, the susceptibility may be represented by the relation $(\chi_A + 20)$ $(T + 90°) = 0.250$. From this, $\mu_A = 1.4$, and, if the spin-only formula may be applied, the number of unpaired electrons is about 0.65. From this it may be concluded that μ_{Pd} is approximately 0.65 Bohr magneton, unless some change occurs at the elevated temperature used.

Rather more convincing evidence is obtainable from magnetic susceptibility data on the palladium-silver alloy system. (The palladium-copper system has been studied but there appears to be some irreversibility which is difficult to interpret. Furthermore, palladium-silver seems to offer a more appropriate analogy, at least for comparison with the nickel-copper system). Svensson[29] and Wucher[28] are in agreement that the susceptibility of palladium drops with increasing silver, and that it becomes zero at approximately 50 atom % of silver, corresponding to $\text{PdAg}_{1.0}$. If the palladium-silver system is similar to nickel-copper and if we may ignore any possible orbital contribution in palladium, then we have only to recall that the moment derived from susceptibility measurements is larger than the saturation by the fraction $2[S(S + 1)]^{1/2}/2S$ to see that the "saturation" moment of pal-

TABLE VII

COLLECTED MAGNETIC DATA ON SYSTEMS RELATED TO NICKEL-HYDROGEN

Atom ratio for $\mu = 0$	Kind of System	$\bar{\mu}_A$	ϵ
$\text{NiCu}_{1.17}$	Substitutional alloy	0.61	0.46
$\text{PdAg}_{1.0}$	Substitutional alloy	$\sim$0.6	$\sim$0.5
$\text{PdH}_{0.64}$?	Interstitial	$\sim$0.6	1.0?
$\text{NiH}_{0.86}$	Adsorption complex	0.61	0.71
$\text{CoH}_{3.4}$	Adsorption complex	1.7	$\sim$0.5

ladium is quite probably very close to that of nickel. If all this is true then we may summarize our information as in Table VII.*

* Dr. A. I. Schindler has kindly pointed out to the author that while the

The state of d electrons in transition metals is still so obscure,[30] and the complications which may occur at metal surfaces so far from being understood, that it would be foolhardy to attempt any categorical statement concerning bond type between metal and adsorbed hydrogen. It may be hoped that sooner or later satisfactory theories of transition metals may be developed and that, when this happens, the magnetic data presented above will prove useful. The data for nickel *seem* to suggest some localization of the bond. They seem to be saying that the nickel atom closest to the hydrogen atom is the only one seriously affected and that it becomes, in effect no longer an active partner in the metal particle ensemble. Adjacent nickel atoms suffer only a trifling peripheral effect. This is in contrast to the palladium-hydrogen system in which the electron from the hydrogen *seems* to be given to the whole adsorbent mass. On the other hand, the data for cobalt[31] show that no such simple explanations are applicable to this metal. Concerning these obscure problems we shall have a little more to say after presentation of the data obtained at low values of M/M_s.*

---

heat capacity study of F. E. Hoare and B. Yates [*Proc. Roy. Soc.* **240**, 42 (1957)] provides good evidence for the existence of about 0.6 electron hole in the palladium d band, yet it is scarcely appropriate to attempt a description of the interstitial system Pd-H in terms of the substitutional alloys Pd-Ag and Ni-Cu. He points out also that Pd-H is actually a two-phase system which may become paramagnetic at low temperature, thus throwing some doubt on the Mott hypothesis concerning the Pd-H system.

In view of the above we may have even more serious reservations about attempts to relate the effect of adsorbed hydrogen on nickel to the other systems shown in Table VII. The writer is indebted to Dr. Schindler for discussion and correspondence concerning this and related problems.

* Some of the considerations given above may possibly aid in understanding the so-called "solubility" of hydrogen in nickel. According to E. C. Stoner's explanation [*Phil. Trans. Roy. Soc. London Ser. A,* **235**, 165 (1936)] the anomalous approach to magnetic saturation in nickel reported by P. Weiss and R. Forrer [*Ann. phys.* **5**, 153 (1926)] may "be accounted for on the basis of a fraction, f_a, of the magnetization being due to a set of [single] domains of comparable size containing an average n_a atoms." Stoner calculates that the anomaly could be caused by 1% of the nickel being present in domains each containing 3000 atoms, that is to say, having an average diameter (if spherical) of about 32 A. A simple calculation will show that if Stoner's single domains were isolated particles then every gram of nickel should be capable of taking up about 0.2 cc of hydrogen by adsorption on the surface of the 1% of metal present in this form. This is much

REFERENCES

1. L. E. Moore and P. W. Selwood, *J. Am. Chem. Soc.* **78**, 697 (1956).
2. R. E. Dietz, Doctoral Dissertation, Northwestern University, Evanston, Illinois, 1960.
3. R. E. Dietz and P. W. Selwood, *J. Chem. Phys.* **35**, 270 (1961).
4. C. P. Bean and J. D. Livingston, *J. Appl. Phys.* **30**, 126S (1959).
5. C. P. Bean, J. D. Livingston, and D. S. Rodbell, *J. phys. radium* **20**, 298 (1959).
6. J. W. Cahn, I. S. Jacobs, and P. E. Lawrence, quoted by C. P. Bean and J. D. Livingston in ref. 4.
7. A. Knappwost and A. Illenberger, *Naturwissenschaften* **45**, 238 (1958).
8. W. Henning and E. Vogt, *J. phys. radium* **20**, 277 (1959).
9. E. Kneller, *Z. Physik* **152**, 574 (1958).
10. C. R. Abeledo and P. W. Selwood, *J. Appl. Phys.* **32**, 229S (1961).
11. G. C. A. Schuit and N. H. de Boer, *Rec. trav. chim.* **70**, 1080 (1951).
12. R. P. Eischens and W. A. Pliskin, *Advances in Catalysis* **10**, 1 (1958).
13. W. A. Pliskin and R. P. Eischens, *Z. physik. Chem. (Frankfurt)* **24**, 11 (1960).
14. P. H. Lewis, *J. Phys. Chem.* **64**, 16 (1960).
15. G.-M. Schwab, J. Block, and D. Schultze, *Angew. Chem.* **71**, 101 (1959).
16. D. Schultze, Inaugural Dissertation, The Ludwig-Maximilian University, Munich, 1958.
17. C. P. Bean, "Structure and Properties of Thin Films," p. 331. Wiley, New York, 1959; *Phys. Rev.* **116**, 1441 (1960).
18. C. A. Neugebauer, *J. Appl. Phys.* **31**, 152S (1960); also personal communication.
19. P. Argyres and C. Kittel, *Acta Met.* **1**, 241 (1953).
20. D. M. S. Bagguley, *Proc. Roy. Soc.* **A228**, 549 (1955).
21. J. Crangle and M. J. C. Martin, *Phil. Mag.* **4**, 1006 (1959).
22. D. A. Dowden and D. Wells, *Actes 2e congr. intern. catalyse, Paris, 1960* **2**, 1499 (1961).
23. D. P. Hollis and P. W. Selwood, *J. Chem. Phys.* **35**, 378 (1961).

more than enough to account for the "solubility" of hydrogen in nickel at room temperature which may be estimated ["Metals Reference Book," 2nd ed., Vol. II., pp. 533–545. Butterworths, London, 1955] as about 0.004 cc/g.

The view expressed above, namely, that the supposed solubility of hydrogen in nickel in the room temperature region is actually an adsorption process on very small crystallites, will be found in D. P. Smith, "Hydrogen in Metals" (pp. 36, 37, and 66. Univ. Chicago Press, Chicago, 1947), where it is attributed to W. A. Wood [*Phil. Mag.* **20**, 964 (1935); *Trans. Faraday Soc.* **31**, 1248 (1935)]. Actually, Wood makes no mention of this matter of hydrogen being adsorbed on the very small crystallites. The writer's impression [supported by personal correspondence] is that H. S. Taylor of Princeton University, was actually the source of this idea some time in the late 1920's or early 1930's.

24. J. J. Broeder, L. L. van Reijen, W. M. H. Sachtler, and G. C. A. Schuit, *Z. Elektrochem.* **60**, 838 (1956).
25. S. A. Ahern, M. J. C. Martin, and W. Sucksmith, *Proc. Roy. Soc.* **A248**, 145 (1958).
26. C. Kittel, "Introduction to Solid State Physics," 2nd ed., p. 330. Wiley, New York, 1956.
27. N. F. Mott and H. Jones, "Theory of the Properties of Metals and Alloys," p. 195. Clarendon Press, Oxford, 1936.
28. J. Wucher, *Ann. phys.* **7**, 317 (1952).
29. B. Svensson, *Ann. Physik* **14**, 699 (1932).
30. C. Herring, *J. Appl. Phys.* **31**, 3S (1960).
31. C. R. Abeledo, Doctoral Dissertation, Northwestern University, Evanston, Illinois, 1961.

CHAPTER VI

The Measurement of Magnetization at Low H/T

However satisfying the direct measurement of saturation magnetizations may be, it cannot be overlooked that few catalytic reactions proceed with measurable velocity at 4.2°K. Surface chemistry will best be served by physical measurements under conditions which favor chemical reactivity. This means that measurements should be made at room temperature and higher, at pressures up to and above 1 atm, and that if the large magnet may be dispensed with, so much the better.

Fortunately, all these conditions may be met. If the results of such measurements cannot be expressed in quite such precise and fundamental terms, they are nevertheless attended with such flexibility, and they yield such a wealth of information, as to make the method almost without parallel in its particular area of application. The method will, in brief, give the number of bonds formed by almost any adsorbate under a very wide range of conditions such as are normally encountered in catalytic practice.

1. The AC Permeameter

Magnetizations on samples exhibiting collective paramagnetism are conveniently found by an induction method, or permeameter, which consists essentially of a step-down transformer. The apparatus is a modification of that described by Heukelom et al.[1]

Design of the gas-handling part of the apparatus, shown diagrammatically in Fig. 31, will, of course, depend on the particular problem at hand. For routine measurements, as of hydrogen on nickel-silica, the apparatus consists of a gas purification train, buret, manometer, sample chamber, McLeod gauges, and vacuum pumps, together with appropriate traps as required. Some modifications, as may be needed for specific problems, will be described later.

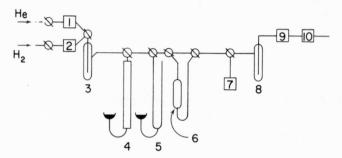

Fig. 31. Gas handling system for low frequency AC permeameter studies: 1, helium purification; 2, hydrogen purification; 3, silica gel trap; 4, gas buret; 5, mercury manometer; 6, sample; 7, McLeod gauge; 8, trap; 9, diffusion pump; 10, mechanical pump.

The only part of the apparatus which may not be familiar to anyone with experience in the area of adsorption is the permeameter itself. This surrounds the sample chamber which may conveniently be a cylindrical glass tube about 10 mm in diameter, 30 mm long, and holding approximately 5 g of pelleted sample. The sample is sealed into the adsorption train prior to reduction and evacuation.

The sample is surrounded by a small solenoid which serves as the secondary of the permeameter. It is convenient to have the secondary made up in two coils arranged coaxially, and several centimeters apart. The sample is placed in the core of one coil and the other, which is connected in opposition to the first, is left empty.* This arrangement makes possible a considerable increase in sensitivity. The secondary assembly is shown in Fig. 32. Each coil of the secondary has about 40 turns. It is quite convenient to construct the secondary so that it may be heated to the reduction temperature, namely, about 400°C. This obviates the difficulty of resetting the sample during the course of any series of measurements.

The wire used for the secondary should be nonmagnetic and not readily oxidized at moderately elevated temperature. Chromel A

* Umeda[2] has recently described a secondary assembly consisting of two sample chambers, to only one of which the adsorbate is added. The secondary is also wound directly on the sample chambers. This arrangement permits an increase of sensitivity but, as designed, it requires a transfer of sample after reduction. The writer is indebted to Dr. Umeda for bringing this to his attention.

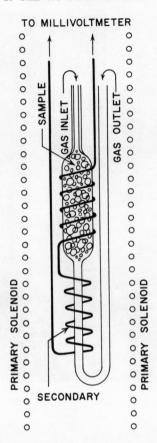

FIG. 32. Arrangement of secondary coils and sample in the low frequency AC permeameter.

No. 26 is satisfactory for this purpose. This is wound on a threaded form machined from "Lava,"* which is then fired to give a strong ceramic-like body, which has given much satisfaction for this purpose. The secondary assembly is mounted on aluminum rods which serve both as supports and as electrical conductors, leading to a Hewlett-Packard Model 400D vacuum-tube millivoltmeter reading down to 0.001 volt full scale. It will be noted that the secondary has quite a low impedance. This is intentionally so to make the millivoltmeter relatively stable to electrical transients. The complete secondary is shown in Fig. 33.

* Available from American Lava Corporation, Chattanooga, Tennessee.

Fɪɢ. 33. Photograph of the secondary coil assembly for the permeameter.

The secondary coils and sample are surrounded by a primary coil which may be raised or lowered as desired. The assembly is shown in Fig. 34. The primary consists of about 5000 turns of Nylform† insulated No. 15 copper wire wound in about 20 layers, some 30 cm long, on a brass core of about 7.5 cm internal diameter. The large core is for introduction of temperature control equipment around the sample.

The primary coil is operated at about 0.75 amp 60 cycle alternating current. The power supply is stabilized by a model 1000-2S Sorenson voltage regulator. If long runs are planned, it is advantageous to cool the primary core by a few turns of copper tubing carrying running water.

The choice of stabilized 60 cycle AC requires some explanation.

† Phelps-Dodge Copper Products Company, New York, New York.

Fig. 34. Photograph of the complete low frequency AC permeameter assembly for obtaining magnetization-volume isotherms.

This apparatus will give an accurate measure of the magnetization of the sample under certain conditions only. The principal condition is that, as discussed on pp. 38–41, the sample particles should have a relaxation time short enough so that the magnetization will reach a maximum in a time short compared with the applied power cycle. A typical nickel-silica catalyst preparation generally fulfills this requirement, but not by a very large margin. It would be impractical to use a higher frequency, but the use of alternating current makes the whole procedure very convenient indeed. Alternatively, it is possible, as pointed out by ,Heukelom *et al.*[1] to use direct current in the primary. This is interrupted, or reversed, and the secondary current induced is integrated in a ballistic galvanometer or fluxmeter. This method has some advantages in that the range of field strength may rise to about 1000 oersteds, as opposed to about 100 oersteds for the AC permeameter. But the other conveniences of the AC method are not easily abandoned.

It is readily shown,[3] as might be expected from the geometry of the apparatus, that the emf produced in the secondary assembly is directly proportional to the magnetization, M, of the sample.* But this is true only, as pointed out above, if the sample exhibits true collective paramagnetism under the experimental conditions. Dietz[3] shows that two tests may conveniently be made to establish this condition. The first is that as M should vary inversely as the absolute temperature and directly as the square of the spontaneous magnetization [Eq. (3.9)], we may write:

$$\frac{E(T_1) - E_0}{E(T_2) - E_0} = \frac{T_2 I_{sp}^2(T_1)}{T_1 I_{sp}^2(T_2)} \tag{6.1}$$

where $E(T_1)$, $E(T_2)$ are the secondary emf readings obtained for a sample at temperatures T_1 and T_2, E_0 is the null emf before insertion (or reduction) of the sample; and $I_{sp}(T_1)$, $I_{sp}(T_2)$ are the spontaneous magnetizations at T_1, T_2, respectively.

Similarly, the emf should be directly proportional to the primary current, whence:

$$\frac{E(i_1) - E_0(i_1)}{E(i_2) - E_0(i_2)} = \frac{i_1}{i_2} \tag{6.2}$$

where i is the current through the primary.

A third possible test relates to the slope of the magnetization-volume isotherm and will be discussed in a later section.

If it is so desired the secondary emf may be recorded directly. In the writer's laboratory a Varian G-10 recorder with full scale sensitivity of 1 mv, and 1-second travel time is used. This is connected through appropriate rectification and an isolating transformer. The recorder is a convenience but it is not actually necessary.

---

* The null emf, prior to insertion of the sample, may be made negligible or nearly so, by adjusting the relative positions of primary and secondary.

Heating of the sample is done with a tube furnace which slips through the core of the primary. For reductions at 400°C, the primary is moved out of position, but the furnace may be used as a thermostat for measurements above room temperature. The furnace is heated with direct current under these conditions. Owing to the relation of magnetization to temperature it is necessary to thermostat the sample to about ±0.1° while data for an isotherm are being obtained. This is done by flowing water from a bath at constant temperature through a cylindrical container surrounding the sample. For measurements below room temperature a Dewar flask fits into the core of the primary. Modifications necessary for work at elevated pressures will be described later.

This section will be concluded with a brief description of general procedure which may be followed for obtaining magnetization-volume isotherms for, say, hydrogen on nickel-silica. The sample in the form of $\frac{1}{8}$-inch pellets is sealed into the sample chamber in such a manner as to fill, as completely as convenient, the core of one of the two opposing secondary solenoids. Five grams of a typical sample containing 40% of nickel, as metal, will give rather more than enough precision.

The sample is reduced in flowing, purified hydrogen at 400°C for about 12 hours, evacuated down to a pressure of about 10^{-6} mm Hg at 400° for 2 hours, and then allowed to cool to the temperature of measurement. If this temperature is much below 0°C, it may be necessary to introduce a trace of helium to hasten the attainment of thermal equilibrium. After the sample has reached the desired temperature the secondary emf is obtained for a fixed primary current. Increments of hydrogen are then admitted to the sample. After each increment it is essential to wait until the sample returns to temperature. At low surface coverages, when the heat of adsorption is still quite high, it may be necessary to wait 10 minutes or more until the heat liberated at each addition of hydrogen is dissipated.

Through successive quantitative additions of hydrogen, after each of which the pressure and the secondary emf are found, there will be obtained data from which may be plotted the magnetization-volume isotherm. The fractional change of magnetization is given by:

$$\frac{\Delta M}{M} = \frac{\Delta(E - E_0)}{E - E_0} \tag{6.3}$$

Pressure-volume isotherms are, of course, obtained simultaneously. It is convenient to plot the final result as the fractional change of magnetization against cc (STP) of hydrogen adsorbed per gram of metal. A virtue of this experimental method is that the correction for gas in the dead-space may, by proper design, be kept quite moderate. Except at elevated pressures the volume of gas in the dead-space may rarely exceed 20% of the total volume of adsorbate admitted to the sample chamber.

2. The Theory of Low Saturation Measurements

If all the particles of a sample exhibiting collective paramagnetism have the same size, as never occurs in practice, then the magnetization at $M/M_s \ll 1$ is given by Eq. (3.9), which may then be rewritten as:

$$M = n_i \frac{(I_{sp}v_i)^2 H}{3kT} \tag{6.4}$$

where as before, n_i is the number of particles of volume v_i in the sample.

It may be wondered why we concern ourselves with an ideal situation which no one has yet achieved, namely, a specimen containing particles of one size only. The reason is that it is easier to understand the derivation for the isotherm if this ideal situation is treated first and, furthermore, it is beginning to appear that real samples behave in part as though they actually do contain such particles.

If hydrogen is adsorbed on a particle of moment $I_{sp}v_i$, the moment will change to become $I_{sp}v_i - n_H \epsilon \beta$ where, as before, n_H is the number of hydrogen atoms so adsorbed, ϵ is the change in the number of Bohr magnetons caused by one hydrogen atom, and β is the Bohr magneton. Hence the magnetization, M', of the sample as thus altered by the presence of adsorbed atoms will be:

$$M' = n_i \frac{(I_{sp}v_i - n_H \epsilon \beta)^2 H}{3kT} \tag{6.5}$$

$$\frac{\Delta M}{M} = \frac{[n_i(I_{sp}v_i - n_H \epsilon \beta)^2 H/3kT] - [n_i(I_{sp}v_i)^2 H/3kT]}{n_i(I_{sp}v_i)^2 H/3kT}$$

$$= \frac{-2n_H \epsilon \beta}{I_{sp}v_i} + \left[\frac{n_H \epsilon \beta}{I_{sp}v_i}\right]^2 \tag{6.6}$$

From Eq. (6.6) we see that the fractional change of magnetization caused by adsorbed molecules varies directly as the number of such molecules taken up by the sample, but is independent (because $v_i n_i$ is a constant) of the volume of the particles. While the temperature does not appear directly in the above expression, it will be recalled that I_{sp} diminishes somewhat with increasing temperature. One consequence of this is that the magnetization-volume isotherm should have a somewhat steeper slope as the temperature is raised.

It will be noted also that the second, squared, term will cause the isotherm to bend away from the volume axis as surface coverage is increased. In practice, $\Delta M/M$ rarely exceeds 20%. A consequence of this is that the isotherm should be approximately a straight line, the slope of which is independent of particle volume. A given mass of smaller particles will, of course, adsorb more hydrogen than will the same mass of larger particles. The magnetization-volume isotherm for a more finely dispersed system will be longer than that for a coarser system but it should have the same slope. A typical theoretical isotherm for a monodispersion is shown in Fig. 35.

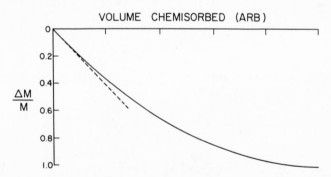

FIG. 35. The theoretical dependence of relative magnetization $(\Delta M/M)$ on volume of adsorbate, for a monodispersed adsorbent (after Dietz).

The derivation of Eq. (6.6) contains two important assumptions in addition to the arbitrary one that all particles are the same size. One assumption is that the adsorbed molecules are uniformly distributed on the surface of all ferromagnetic particles. The other is that successive increments of hydrogen have the same effect on the magnetization or, to put it another way, that ϵ does not change with increasing surface coverage, or particle size. Both of the assumptions lie at the heart of any treatment of chemisorption. Both will be dealt with in detail in the next chapter.

Next we shall present the treatment developed by Dietz[3] for samples containing a distribution of particle diameters. From Eq. (6.5) we see that:

$$M' = \frac{H}{3kT} \sum_i n_i(\mu_i - \Delta\mu_i)^2 \tag{6.7}$$

where μ_i is the moment of particles of radius r_i, and $\Delta\mu_i$ is the change of moment caused by the adsorbate.

The fractional change in magnetization is then:

$$\frac{\Delta M}{M} = \frac{\sum n_i(\Delta\mu_i)^2 - 2\sum n_i\mu_i\Delta\mu_i}{\sum n_i\mu_i} \tag{6.8}$$

If the particles under consideration are spherical, then:

$$\Delta\mu_i = 4\pi r_i^2 s_H \theta\epsilon\beta \tag{6.9}$$

where s_H is the number of hydrogen atoms adsorbable per unit area, and θ is the fractional surface coverage.

It will be noted that:

$$\frac{(s_H\theta\epsilon\beta)A_p}{I_{sp}V} = \frac{\Delta M_s}{M_s} \tag{6.10}$$

Where A_p is the total surface area of the particle, whence, substituting for $s_H\theta\epsilon\beta$ in Eq. (6.9):

$$\Delta\mu_i = \frac{\Delta M_s}{M_s}\left(\frac{I_{sp}V}{A_p}\right)4\pi r_i^2$$

$$= \frac{\Delta M_s}{M_s}\left(\frac{I_{sp}V}{A_p}\right)4\pi\left(\frac{3v_i}{4\pi}\right)^{2/3} \tag{6.11}$$

where v_i, as above, is the volume of a particle of radius r_i.

We transform n_i into a continuous volume distribution function:

$$n_i = \frac{f(v)dv}{v} \tag{6.12}$$

where:

$$\int_0^\infty f(v)dv = V \tag{6.13}$$

Furthermore, expressing A_p in terms of the continuous distribution function:

$$A_p = \int_0^\infty 4\pi r^2 \frac{f(v)dv}{v}$$

$$= 4\pi \left(\frac{3}{4\pi}\right)^{2/3} \int_0^\infty \frac{f(v)dv}{v^{1/3}} \tag{6.14}$$

and substituting into Eq. (6.8), we have:

$$\frac{\Delta M}{M} = - \alpha \left(\frac{\Delta M_s}{M_s}\right) + \delta \left(\frac{\Delta M}{M}\right)^2 \tag{6.15}$$

where

$$\alpha = 2 \frac{\int_0^\infty v^{2/3}f(v)dv \int_0^\infty f(v)dv}{\int_0^\infty v^{-1/3}f(v)dv \int_0^\infty vf(v)dv} \tag{6.16}$$

and

$$\frac{\alpha}{\delta} = 2 \frac{\int_0^\infty v^{2/3}f(v)dv \int_0^\infty v^{-1/3}f(v)dv}{\int_0^\infty f(v)dv \int_0^\infty v^{1/3}f(v)dv} \tag{6.17}$$

The symmetry of the ratio of integrals suggests that both α/δ and α are both approximately equal to 2. Experimental results obtained in this writer's laboratory confirm this view.

Table VIII shows values of α and α/δ calculated for certain arbitrary particle size distributions.[3] It is, therefore, clear that provided the sample exhibits true collective paramagnetism the low field method gives a change $\Delta M/M$, which is proportional to the change of saturation magnetization. We note also that the

TABLE VIII

THE QUANTITIES α AND α/δ, EQ. (6.15), FOR CERTAIN ARBITRARY
DISTRIBUTIONS OF PARTICLE SIZE[a]

Distribution	α	α/δ
Delta function	2.000	2.000
Infinitely wide rectangle	1.600	2.400
Maxwellian (volumes)	1.867	2.100
Maxwellian (radii)	1.471	2.500

[a] After Dietz.[3]

second, squared, term may be neglected if $\Delta M/M$ is not large. Equation (6.15) reduces to Eq. (6.6) for a sample in which all the ferromagnetic particles are the same size. This is true because, as $n_H = s_H\theta A_p$, it follows that $\Delta M_s/M_s = -n_H\epsilon\beta/I_{sp}v$.

Finally, in this section, we shall consider the case of a sample in which there are two sizes of particles, and in which these particles are all accessible (though not necessarily at the same rate) to vapor molecules. If this assumption is not always met in practice we shall, nevertheless, find it to be a useful exercise.

From Eq. (6.6), $\Delta M/M = -2n_H\epsilon\beta/I_{sp}v$, one sees that for two assemblies of particles of volumes v_1 and v_2, we may write:

$$\frac{\Delta M_1/M_1}{\Delta M_2/M_2} = \frac{-2(n_H)_1\epsilon\beta/I_{sp}v_1}{-2(n_H)_2\epsilon\beta/I_{sp}v_2}$$

$$= \frac{(n_H)_1v_2}{(n_H)_2v_1} \tag{6.18}$$

where $(n_H)_1$, $(n_H)_2$ are the numbers of hydrogen atoms adsorbed per particle of volumes v_1, v_2, respectively.

We find also from Eq. (6.4), $M = n_i(I_{sp}v_i)^2H/3kT$, that

$$\frac{M_1}{M_2} = \frac{n_1v_1^2}{n_2v_2^2} \tag{6.19}$$

so that, substituting for M_1, M_2 in Eq. (6.18):

$$\frac{\Delta M_1}{\Delta M_2} = \frac{(n_H)_1n_1v_1}{(n_H)_2n_2v_2} \tag{6.20}$$

n_1, n_2 being, as before, the total number of particles of volumes v_1, v_2, respectively.

One sees, therefore, that we may relate the quantities of adsorbent present in each of the two size categories to the change of magnetization caused by adsorbed hydrogen on each category. This, as will be shown in the next chapter, is not quite so nearly impossible as it may at first appear.

3. Thermal Transients and the Heat of Adsorption

The abrupt admission of hydrogen to a reduced, evacuated nickel-silica sample always results in magnetization changes as shown in Fig. 36. These consist of a rapid decrease of magnetization followed by a more leisurely partial recovery. This effect is

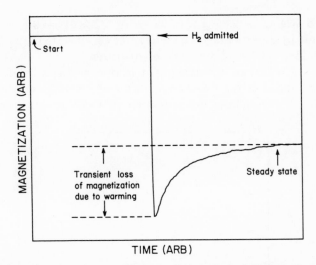

FIG. 36. Thermal transient observed when hydrogen is abruptly admitted to an evacuated nickel-silica sample.

attributed to warming of the nickel particles through liberation of the heat of adsorption, followed by a return to the ambient by the hydrogen-covered particles. There is a possibility, however, that part of the transient effect may be due to hydrogen moving from more to less accessible sites situated on much smaller particles.

This phenomenon of thermal transients may be used to estimate the heat of adsorption. In the first paper[4] in which reference was made to this effect the change of magnetization with temperature, as determined by the Faraday method, was used to find the temperature rise involved. A more realistic procedure[3] is to consider that for a sample exhibiting collective paramagnetism (the only ones to which the method is applicable) the magnetization must vary inversely as the absolute temperature. A sample of catalyst weighing 5.41 g and containing 52.8% of nickel as metal was abruptly flushed with hydrogen to 1 atm pressure after having previously been evacuated at 400°C. The temperature of adsorption was 20°C. It was found that the magnetization quickly fell until $\Delta M/M = 0.29$, but then came to equilibrium at $\Delta M/M = 0.184$. The total sorption of hydrogen was 45.6 cc (STP).

The excess, transient loss of magnetization was 11%, which could have been caused by a temporary rise in temperature of about 31°. The heat necessary to raise 2.85 g of nickel metal from

293° to 324°C is 2.85 g $\times$ 0.105 cal/g/°C $\times$ 31° = 9.1 cal. One mole of hydrogen would therefore, have liberated $(22.4 \times 10^3 \times 9.1)/45.6 = 4.5$ kcal per mole of hydrogen.

This is, of course, the integrated heat over the whole surface coverage and it is too small by a factor of about 3, as compared with direct calorimetric determination. We do not have any ac-

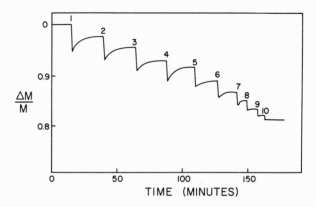

FIG. 37. Diminishing thermal transients observed with increasing surface coverage of nickel by hydrogen (after Lee).

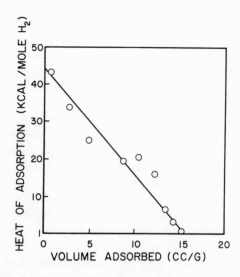

FIG. 38. Incremental heat of adsorption for hydrogen on nickel calculated from the data shown in Fig. 37 (after Lee).

curate knowledge concerning the maximum temperature reached by the nickel, and we do not know how quickly the heat is distributed to the silica support. Under the circumstances it is about all that could be expected that the heat is about the correct order of magnitude.

A more useful procedure is to use the method for comparing heats of adsorption at different levels of surface coverage. The change in size of the thermal transients with successive increments of adsorbate is shown in Fig. 37, and from this it is possible to construct a plot of differential heats.[5] Such a plot is shown in Fig. 38. While the absolute accuracy is quite poor, there is no doubt that the decreasing heat with increasing coverage is clearly reflected by this method which has the unique feature that the nickel is caused to act as its own thermometer.

REFERENCES

1. W. Heukelom, J. J. Broeder, and L. L. van Reijen, *J. chim. phys.* **51**, 474 (1954).
2. S. Umeda, *J. Japan Soc. Powder Met.* **8**, 159 (1961).
3. R. E. Dietz, Doctoral Dissertation, Northwestern University, Evanston, Illinois, 1960, p. 111.
4. P. W. Selwood, *J. Am. Chem. Soc.* **78**, 3893 (1956).
5. E. L. Lee, J. A. Sabatka, and P. W. Selwood, *J. Am. Chem. Soc.* **79**, 5391 (1957).

CHAPTER VII

Magnetization-Volume Isotherms for Hydrogen

1. Experimental Data for Hydrogen on Nickel

A typical plot of fractional magnetization change, $\Delta M/M$, against volume of hydrogen adsorbed per gram of nickel present, is shown in Fig. 39. This plot was made from data obtained on a commer-

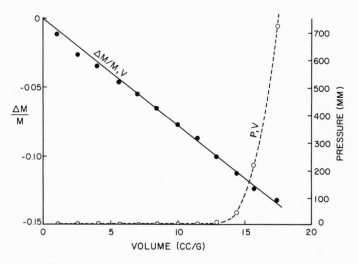

Fig. 39. Magnetization-volume and corresponding pressure-volume isotherms obtained at room temperature for hydrogen adsorbed on nickel.

cial nickel-kieselguhr containing about 50% of nickel, reduced and evacuated as described in the last chapter. The data refer to room temperature and, for convenience, the corresponding pressure-volume isotherm is also shown. Various catalyst preparations obtained through many different procedures and containing quite different proportions of nickel yield similar magnetization-volume isotherms.[1] The one essential is that all preparations must pass the tests (p. 94) for collective paramagnetism. In view of this virtual universality in general form of the isotherm, we shall devote considerable attention to its properties and to what it can tell us about the mechanism of chemisorption. The present section will be concerned with the experimental properties of the isotherm.

If, after the data for plotting a magnetization-volume isotherm are obtained, the sample is evacuated, it will be found that at room temperature approximately one-third of the hydrogen may be desorbed. Prolonged evacuation removes a negligible additional volume of adsorbate. Within this limited region the magnetization is recovered linearly with the hydrogen; the isotherm is, therefore, strictly reversible throughout the region in which hydrogen may be pumped off. Careful studies[2] have shown that all the hydrogen originally admitted may, within experimental error, be recovered by evacuation for about 2 hours at 360°C. If, after this treatment, the sample is cooled to room temperature, it will generally be found that the magnetization is slightly greater than it was at the same temperature before admission of the hydrogen. The reason for this increase, which occurs progressively with every adsorption-desorption cycle, seems to be that the nickel particles slowly increase in size. But whether this is due to the heating necessary to desorb the hydrogen, or to some irreversible change related to the presence of the hydrogen, is not known. This effect concerns us only, as will be shown in the next chapter, when it becomes necessary to compare the isotherm slope for hydrogen with that of some other adsorbate.

We shall first consider whether the slope of the isotherm agrees with the prediction of Eq. (6.6), with due regard for the fact that real samples exhibit a considerable range of particle size.

Neglecting the squared term, we may write Eq. (6.6) as follows:

$$\frac{\Delta M}{M} = \frac{-2n_{\mathrm{H}}\epsilon\beta}{I_{\mathrm{sp}}v_i} \tag{7.1}$$

It will be worthwhile to review the meaning of this equation which states that the fractional loss of magnetization caused by adsorbed hydrogen is dependent directly on the number, n_{H}, of hydrogen atoms taken up per particle of adsorbent, and inversely as the particle volume, v_i. If numerator and denominator are multiplied by n_i, the number of particles present per unit volume of ferromagnetic adsorbent, we have:

$$\frac{\Delta M}{M} = \frac{-2N_H\epsilon\beta}{I_{sp}V} \tag{7.2}$$

where N_H is the number of hydrogen atoms adsorbed on the sample and V is the volume of ferromagnetic adsorbent. Then for a typical nickel-silica adsorbing say 10 cc (STP) H_2 per gram Ni at 25°C, we have:

$$\frac{\Delta M}{M} =$$

$$\frac{-2 \times [2 \times 6.02 \times 10^{23} \times (10/22.4 \times 10^3)] \times (0.72) \times (9.27 \times 10^{-21})}{(4.85 \times 10^2) \times (1/8.9)}$$

$$= -0.131 \text{ (calc)}$$

This value may be compared with experimental results as follows: Selwood,[3]—0.080; Lee et al.,[4]—0.105; Broeder et al.,[5]—0.290.

While the above results are certainly within the correct order of magnitude, they leave something to be desired. The reason for the poor agreement is readily found in the assumption that all particles of metal in the sample have the same volume. It will be recalled (p. 99) that if we use the more accurate expression Eq. (6.15), it is found that the constant α becomes equal to 2, as given in Eq. (7.2), only for uniform particles. The range of α shown in Table VIII, with various assumed particle size distributions, is more than sufficient to account for the variation of slope actually found for the magnetization-volume isotherm.

Equation (7.2) shows that the slope of the isotherm should vary inversely as the spontaneous magnetization. Inasmuch as the spontaneous magnetization is a function of temperature, we may expect the isotherm slope to increase moderately with increasing temperature, even in the room temperature region. Reference to Fig. 8 will show that I_{sp} decreases slowly, then more rapidly, as the temperature is raised. It finally becomes zero above the Curie point. (This change should not be confused with the fact that the magnetization M varies inversely as the absolute temperature for any sample exhibiting collective paramagnetism.)

Figure 40 shows isotherms[6] for hydrogen on nickel-silica at —78°C and at 206°. These isotherms were obtained under conditions as nearly identical as possible except for the temperature

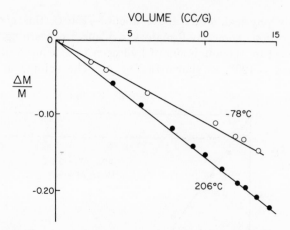

FIG. 40. Magnetization-volume isotherms at two different temperatures for hydrogen on nickel.

change. The ratio of isotherm slopes is 0.71, increasing with increasing temperature as predicted. The inverse ratio of spontaneous magnetizations at these two temperatures (from Fig. 8) is 0.69. It will, however, be recalled that very small particles of nickel have a Curie temperature a little lower than that (358°C) of pure massive nickel. For particles of the size under consideration, the difference is about 50° and this results in a corrected magnetization curve for which I_{sp} (206°C)/I_{sp} (−78°C) = 0.73, which is very satisfactory agreement.* The most significant feature of this agreement is, however, that it establishes the invariance of ϵ over a wide temperature range. From −78° to 206°C there is no change in the number of electron spins paired per atom of hydrogen adsorbed. In other words, if we may accept ϵ as a measure of adsorption bond type, there is no change in the mechanism of adsorption of hydrogen on nickel over nearly a 300° range of temperature.

It is well known that the mode of adsorption of hydrogen on nickel is nevertheless, to a degree, dependent on the temperature of adsorption.[7] Thus, at quite low temperatures the mechanism is

* In an earlier paper[3] it was reported that the isotherm slope is *greater* if the magnetization, after admission of hydrogen at 25°C, is measured at −196°. This result obtained by the AC permeameter method is not reproducible by the static, high-field method. The discrepancy may probably be attributed to deviations from collective paramagnetism.

primarily physical, while above about $-150°C$, this goes over to chemisorption. Some confirmatory evidence[3] concerning this view is obtained by the admission of hydrogen to a reduced nickel-silica sample at $-196°$, as shown in Fig. 41. The volume of hydrogen

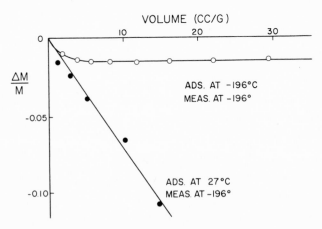

FIG. 41. Magnetization-volume isotherms for hydrogen on nickel showing dependence of adsorption mechanism on temperature of adsorption.

adsorbed is very large, but the change of magnetization is negligible after the first few cubic centimeters of hydrogen have been taken up. This result is in agreement with the conclusion reached by Beeck *et al.* with respect to the adsorption of hydrogen on nickel films at low temperatures. It is also in agreement with the observations of Sadek and (H. S.) Taylor[8] who showed that, on a catalyst similar to that used in this investigation, of the 8.17 cc of H_2 adsorbed (with ascending temperature) per gram Ni, 7.70 cc was physically adsorbed and only 0.47 cc chemisorbed. With descending temperature the respective volumes were approximately equal. This view receives further confirmation from the observation by Schuit *et al.*[9] (confirmed in the author's laboratory) that very little hydrogen-deuterium exchange occurs on nickel-silica at $-196°C$.

It is to be noted[3] that if the sample which holds a rather large volume of van der Waals hydrogen at $-196°$ is allowed to warm to room temperature, there occurs an abrupt loss of magnetization as the transition occurs from van der Waals to chemisorbed hydrogen. These results are, of course, in excellent agreement with those previously reached by various workers from several points of view. The abrupt change of bonding is caused by the self-accelerating effect of the heat of chemisorption on this transition.

It is sometimes stated that the surface coverage of nickel by hydrogen is virtually complete at a fraction of a millimeter pressure. This statement is susceptible of examination because the magnetic method gives an unequivocal test as to whether or not chemisorption has occurred. The experimental method[10] for making this investigation involves somewhat elevated pressures and consequently somewhat different design of the sample holder and buret system. The sample holder consists of a stainless steel, non-magnetic bomb around the outside of which is wound the secondary coil of the permeameter. This design, shown in Fig. 42, results in some loss of sensitivity, but not enough to be serious. The gas buret system is shown in Fig. 43.

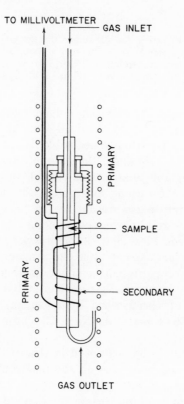

FIG. 42. Modification of permeameter secondary and sample container for obtaining magnetization-volume isotherms at elevated pressures (after Vaska).

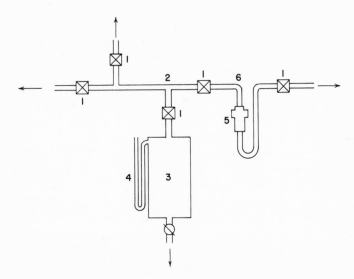

FIG. 43. Gas buret system for obtaining magnetization-volume isotherms at elevated pressures (after Vaska). (1) steel valves, (2) steel pressure tubing, (3) glass exhaust chamber serving as buret, (4) open-arm manometer, (5) adsorbent chamber, (6) stainless steel tubing 1/16 in. i.d.

Figure 44 shows a magnetization-volume isotherm for hydrogen on nickel-silica carried up to about 70 atm. At the same pressure the effect of helium is virtually negligible. When the results shown in Fig. 44 were first obtained, the effect of the squared term in Eq. (6.6) was not appreciated. One should, therefore, attribute a moderate part of the turning away from the volume axis to the squared term. Larger parts are due to the "slow" sorption of hydrogen which, as is well known, always occurs in such systems, and which tends to become less slow as the pressure is raised. But these various considerations do not alter the primary conclusion which is that surface coverage is not complete on this system at 1 atm pressure. Coverage is essentially complete at 100 atm—in a typical run the volume of chemisorbed hydrogen (as determined magnetically) was 17.6 cc per gram Ni at 23.4 atm and 17.9 cc at 71 atm.

One further point to be mentioned in this section is the possible effect of preadsorbed molecules on the magnetization-volume isotherm for hydrogen. In later chapters we shall make frequent use of the technique of blocking certain fractions of the metal surface

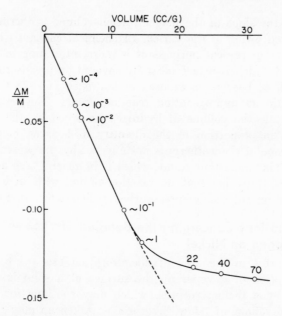

Fig. 44. Magnetization-volume isotherm at room temperature for hydrogen on nickel at pressures, as indicated, up to 70 atm. The dotted line represents the high-pressure part of the isotherm after appropriate corrections (after Vaska).

area by preadsorption. If the preadsorbed molecule is one such as ethylene (at room temperature or higher), then the hydrogen up-

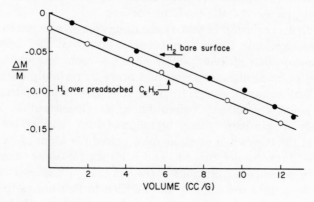

Fig. 45. Magnetization-volume isotherms for hydrogen on nickel at $-78°C$ before and after the preadsorption of cyclohexene (after Den Besten).

take and the slope of the magnetization-volume isotherm may be complicated owing to the variety of chemical changes which may occur. For our present purposes it is more interesting to see what happens if a nickel surface partially covered by preadsorbed molecules such as benzene is exposed to hydrogen at −78°C. At this temperature no hydrogenation reaction occurs. The result found is that while the volume of hydrogen adsorbed by the nickel is decreased in proportion to the quantity of benzene preadsorbed, yet the slope of the isotherm is not affected by the presence of the benzene. This important result, which is in general true of all preadsorbates (provided that no reaction occurs with hydrogen), is illustrated for the case of preadsorbed cyclohexene[11] in Fig. 45.

2. Conclusions Concerning the Bonding of Hydrogen on Nickel

All workers in the area of chemisorption are aware that the rapid sorption of a vapor on the surface of a solid is generally followed by a further sorption which may continue for hours or days. The volume of "slow" hydrogen so taken up may be an appreciable fraction of the whole; and the process has received an amount of attention which is, perhaps, out of proportion to its intrinsic importance. It will be shown in the following section that the slow process is probably only a special case of that which takes place rapidly. This then seems an appropriate place to summarize our observations with respect to hydrogen on nickel and to draw what conclusions we may concerning the nature of the adsorption process and the mode of bonding.

Our first conclusion is that if chemisorption is adsorption which involves electronic interaction between adsorbent and adsorbate, then the sorption of hydrogen by nickel at room temperature and for a substantial distance above and below, is certainly chemisorption. There seems no escape from the conclusion that electron spins in the nickel are actually paired during the chemisorption process. The rate of spin-pairing may be followed with considerable precision and the number of electron spins paired per atom of hydrogen adsorbed may be determined. The evidence for these statements has been presented in previous pages and it is just as conclusive as the evidence that reduction of the Fe^{3+} ion to Fe^{2+} lowers the magnetic moment by approximately one electron spin, or that formation of a cyanide complex lowers the moment of Fe^{3+} from 5.9

Bohr magnetons to 1.7 in Fe $(CN)_6^{3-}$. It is, of course, true that few workers in the field of surface chemistry would dispute the statements made above. The value of the magnetic method is that we may state in quantitative terms what number of electrons has been affected, and this may be done over a diversity of conditions which cover nearly every circumstance encountered in the practical applications of catalysis.

Our next conclusion is that over the greater part of the surface coverage, and over a substantial range of temperature, there is no important change of bond type.* The fact that the magnetization-volume isotherm for hydrogen on a bare nickel surface is very near to the line predicted by Eq. (6.6) is virtual proof that no change of adsorbed species (from, say, H^+ to H_2^+) occurs. In view of the divergent views on this subject, and the wealth of seemingly conflicting experimental data, the subject will be examined in some detail.

The various lines of evidence which suggest that some change of bond type may occur with increasing coverage of nickel by hydrogen are as follows.

The heat of adsorption decreases from about 30 kcal per mole of hydrogen down to 5 kcal or less. This change may possibly be due to increasing formation of chemisorbed species such as H_2^+, or it may be due to a weakly adsorbed layer of hydrogen atoms (or molecules) on top of the monolayer. But the change of heat of adsorption may be due to causes other than a definite change of bond type.[13] It may, for instance, be due to the mutual repulsion of surface dipoles, it may be due to a decrease in work function, it may be due to a change of the Fermi level, or it may be related to some redistribution of the residual metallic bonds. To each of these there may be developed some negative arguments, and it should be mentioned that the concept of changing bond type may very well

---

* Umeda[12] has recently reported that in one of two samples of coprecipitated nickel-silica studied, the slope of the magnetization-volume isotherm increased rather abruptly at about ⅔ of the total sorption, which was about 27 cc per gram of nickel. It is impossible at this writing to know what weight should be placed on this observation. Umeda attributes it to a primary stage of strong chemisorption on smaller particles followed by a weaker stage on larger particles. The idea of uniform distribution of hydrogen per unit of metal surface is implicit in our derivation. If Umeda's observation is correct it will necessitate some reconsideration of the conclusions presented here.

be significant in cases such as that of carbon monoxide on nickel where such a change is acknowledged, even if the precise nature of the change remains obscure. To this writer the most convincing evidence concerning the cause for diminishing heat of adsorption with coverage is that adduced by Gomer,[14] namely, the heterogeneity of the adsorption sites. This old concept receives powerful support from field-emission microscopy. In this connection it may be pointed out that the silica-supported metal particles with which we deal in these magnetic studies are from 10 to 100 times smaller than the tips used in field-emission microscopy. The surface of such small particles must provide an extraordinary opportunity for heterogeneity to develop. This view concerning the change of heat of adsorption is, of course, consistent with the magnetic data presented above.

It may perhaps be questioned whether the magnetic method is sufficiently sensitive to detect changes of bond type which do not involve a change in the relative numbers of adsorbent-adsorbate atoms involved. To this question we can scarcely give an unequivocal answer. It would seem that if electrons from hydrogen are given to the d band of the metal, then ionic bonding of the type Ni^-—H^+ would have the same magnetic affect as a covalent bond. By analogy to what we know about transition metal ions in other than metallic environments, it would appear that the bond type Ni^+—H^- would lead to an increase of magnetic moment. At least this is certainly the case with ions such as Fe^{2+} which contain more than half-filled d orbitals. These, on oxidation, almost invariably suffer an increase of moment. Be that as it may, we may almost certainly say that the magnetic method will detect a change of Ni^-—H^+ to Ni^-—H_2^+, or to Ni_2^-—H^+.

The other lines of evidence which have from time to time, been presented as support for the "change of bond type" hypothesis will give us less trouble. One of these is the electrical conductivity.[15] The conductivity of nickel films prepared by the evaporation technique is changed by the presence of adsorbed gases of which hydrogen is one. The change of conductivity depends on the thoroughness with which the film has been evacuated. Sachtler[16] has shown that the sign of the effect may actually be reversed by evacuation of the sample under rigorous conditions. This conclusion has received quite general confirmation and seems to dispose of earlier arguments based on the supposed change of sign of the conductivity effect produced by hydrogen at different levels of surface coverage.

On the other hand, the change of conductivity appears not to be linear with surface coverage. Sachtler and Dorgelo[17] have obtained data of the kind illustrated in Fig. 46. It will be noted that quite

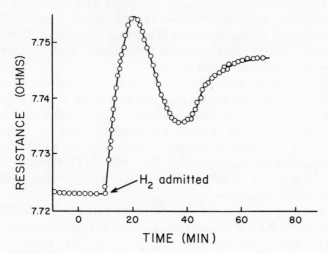

FIG. 46. Electrical resistance of a nickel film versus hydrogen coverage (after Sachtler and Dorgelo).

striking changes of conductivity occur as surface coverage is increased. These authors interpret these data as follows: after the early stages of adsorption by dissociated atoms (giving an increase of resistance), and after the heat of adsorption has been thus diminished, a second molecular hydrogen species, more mobile than the first, causes a decrease of resistance. This second layer may then migrate to the surface of the metal, causing a second increase of resistance.

There may be less contradiction between these conclusions and those based on the magnetic data than at first appears. According to the views presented in previous chapters, Sachtler and Dorgelo's second hydrogen species is not chemisorbed and thus could scarcely be expected to cause electronic spin pairing. Nevertheless, our chief objection to such interpretations is that there is no quantitative relationship between conductivity in such systems and the number of electrons (or atoms) that might be involved in adsorbent-adsorbate bond formation. Attempts to develop such a relationship are discussed by Gundry and Tompkins[13] who seem to share this writer's view that such attempts have, as yet, a very tenuous basis. In a later section it will be shown how the change of magnetization during chemisorption of hydrogen on nickel may also, under certain conditions, be nonlinear with surface coverage (or even show a change of sign). It will also be shown how this observation led to the erroneous belief that a change of bond type had been observed, and how the phenomenon may be shown to have quite a different explanation.

The same conclusion must be applied to arguments based on surface potentials. Before we may say that a set of data reflects a change of bond type, we must establish a quantitative relationship (an equation of state) between the property being measured and the state of the atoms or electrons involved in the bond. This has been done for the magnetization [Eq. (6.5)]. To the best of the writer's knowledge it has not been done for any other property related to the adsorbent.

It would be gratifying if we could, at this point, amplify our previous remarks concerning the nature of the chemisorptive bond (p. 112). Unfortunately, little may be added. The bond between nickel and hydrogen appears to involve at least partial transfer of the electron from the hydrogen to the d band (or d orbital) of the nickel. There is some suggestion that the bond is localized on the particular atom attached. But the data on cobalt show that this simple picture is far from adequate. An alternative suggestion[18] is that the adsorbed atoms create a set of levels which the electrons may fill with the restriction that only states having an energy less than the Fermi energy may be occupied. If the number of levels below E_F is less than half the number of atoms adsorbed (two electrons may occupy each level), the remaining electrons will occupy holes in the $3d$ and $4s$ bands and lower the magnetization. The fact that ϵ for hydrogen on cobalt is so much less than the saturation moment for cobalt ($\mu_A = 1.7\beta$) suggests that the similarity between μ_A and ϵ for nickel should be disregarded and that the explanation invoking the Fermi level is the more probable.[19] This is also not inconsistent with the theoretical treatment of the problem by Grimley,[20] and it does not necessarily invalidate the idea of a localized or covalent bond.

In this connection it is appropriate to refer again to the attempt of Broeder *et al.*[5] to interpret their magnetic data in terms of bond type. Their procedure is as follows:

Each nickel atom has a moment of about 0.6 Bohr magneton, and each nickel particle a saturation moment of 0.6 multiplied by the number of atoms in the particle. The ratio of hydrogen atoms adsorbed at complete coverage to total nickel atoms present is assumed to give the fraction of nickel atoms on the surface, and these nickel atoms are assumed to lose 0.6 Bohr magneton each. It is, therefore, possible to calculate the loss, $\Delta M_s/M_s$, in saturation magnetization for any given sorption of hydrogen and to compare the result with that found experimentally. Any deviation is thought to indicate some deviation (positive or negative as the case may be) from true homopolar bonding. Thus a given nickel-silica sample took up 50.6 cc of hydrogen per gram of nickel, the calculated loss of saturation magnetization is then 0.44.* The observed loss was 0.22. This deviation is stated to establish some diminution in the M^+—X^- structure contribution to the bond resonance hybridization, although to the writer the opposite conclusion would seem to be equally probable.

Be this as it may, our chief objection to the procedure is that the saturation magnetizations were not measured directly, but were merely estimated from measurements at room temperature. It has already been shown (p. 64) that this procedure does not yield the true saturation magnetization. It should be noted also that for such a large change of magnetization it is not possible to ignore the squared term [Eq. (6.6)], and there is the further problem of possible deviations from true collective paramagnetism. In view of all these difficulties it is doubtful if much reliance may be placed on these calculations. This does not, however, exclude the possibility that such effects may yet be discovered.

We may now summarize the several conclusions to be drawn from magnetic data. These are as follows:

1. Electronic interaction is definitely proved to occur for the adsorption of hydrogen on nickel.

2. After the hydrogen has been adsorbed the slope of the magnetization-volume isotherm becomes a little steeper with rising temperature. This effect is due to the decrease of spontaneous

---

* The authors give a calculated loss of 0.52 but this seems to be an arithmetical error. This error in itself does not invalidate the argument.

magnetization, corrected for the small but definite decrease of Curie temperature.

3. Surface coverage is nearly, but not quite, complete at a pressure of 100 atm of hydrogen.

4. The mode of adsorption is independent within fairly wide limits of temperature change after adsorption.

5. The mode of adsorption appears to be independent of the temperature of adsorption down to somewhat below −100°C. Below this temperature physical adsorption becomes increasingly important.

6. Preadsorbed molecules have no effect on the mode of adsorption of hydrogen provided that these molecules do not react chemically with hydrogen.

It should be cautioned that some of the conclusions stated above do not necessarily apply to adsorbates other than hydrogen.

3. The "Slow" Sorption of Hydrogen on Nickel

The slow sorption of hydrogen which generally follows the rapid, nonactivated chemisorption has received attention from several points of view. The view that the process is simply a "solution of hydrogen in the metal" may be dismissed because the volumes of hydrogen taken up by the slow process are much greater than the maximum "solubility" of hydrogen in the adsorbents under consideration.[21] Several other treatments will be discussed in some detail.

The report of Sadek and Taylor[8] is representative of the treatment developed by (H. S.) Taylor over many years, namely, that the phenomenon represents the requirement of an activation energy during formation of the bond. For our purposes this paper is important because the catalyst sample is the familiar nickel-silica. The equilibrium volume of hydrogen sorbed at several temperatures (all below room temperature) was measured and, from the rate of attainment of equilibrium, calculations were made in the usual way to determine the apparent activation energy of the process. Appropriate corrections were made for the not inconsiderable van der Waals' adsorption which occurred at the lower end of the temperature range covered. A typical example of the results obtained is shown in Fig. 47.

The persistence of Taylor's views concerning activated adsorp-

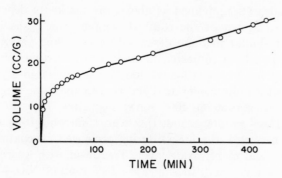

FIG. 47. Rate of adsorption at −128°C for hydrogen on nickel (after Sadek and Taylor).

tion is clear evidence of how difficult it is to prove, or disprove, any interpretation of accepted facts in surface chemistry. Since the theory was first advanced over 30 years ago[22] there have been many attempts to overthrow it. We shall not be able to overthrow it here, but we shall show that another explanation for the slow process has now a substantial accumulation of evidence in its favor. The theory of activated adsorption, so far as it applies to the nickel-hydrogen system, has no explanation for the fact that the slow process continues at temperatures considerably more elevated than the maximum used by Sadek and Taylor, but that the rate of adsorption does not increase, and may even decrease[23] in the neighborhood of 300°C. Furthermore, the rather long evacuation times used by Sadek and Taylor suggest that the Schuit and De Boer oxide film theory (to be discussed below) may have contributed to the "slow" process.

Schuit and De Boer[24] define the slow process in the following manner: First the volume taken up at −196°C and at low pressure is determined. This is designated Type I adsorption. The temperature is then raised to −78° and the pressure to 100 mm. During this latter process a substantial further sorption of hydrogen occurs. This is designated Type II and is considered to be essentially "slow" adsorption. One is, of course, at liberty to define adsorption processes in any way one may see fit. But it should be pointed out that this definition is not precisely the same as that adopted by many other workers in the field.

Schuit and De Boer show that the relative proportions of Types

I and II adsorption, defined as above, are markedly dependent on the prior treatment of the catalyst sample, which was a nickel-silica quite similar to those described on preceding pages. A sample which had been subjected to prolonged reduction followed by a relatively brief evacuation at 400°C shows a large Type I adsorption but a quite moderate Type II. But the same sample after prolonged evacuation at 400° shows negligible Type I but large Type II. These results suggest that water emerging from the silica support reacts with the nickel during prolonged evacuation, forming a superficial nickel oxide layer. Treatment with hydrogen then reduces this layer and, inasmuch as the reaction $NiO + H_2 \rightarrow Ni + H_2O$ requires an activation energy, we find the "sorption" to be increasingly "slow" under conditions where the oxide film has time to grow. The view thus expressed is supported by similar observations on samples which have been exposed to oxygen.

If we have any serious reservations concerning these ideas they are based on the fact that the total volume of hydrogen sorbed is roughly the same for samples in which the relative proportions of Types I and II are markedly different. Yet it might be thought that at room temperature and lower the process could never be that indicated above, but that the reaction leading to Type II would invariably include a substantial volume of hydrogen bonded to the reduced nickel; that is to say, at least one hydrogen atom would be required for every site evacuated by oxygen. In such a case the total volume of hydrogen taken up would be considerably greater whenever Type II predominated, but such appears not to be the case unless, as indicated below, there are some compensating factors such as particle growth to be considered.*

The "slow" process explanation for hydrogen on nickel-silica developed by Liebowitz et al.[25] is based on the Taylor-Thon[26] mechanism of chemisorption which is related to the creation and spontaneous decay of adsorption sites. The rate of the slow process is thought to be determined by the rate of site decay. The data were obtained on a nickel-kieselguhr containing 11.5% of nickel and treated in a manner much like that of the other investigations mentioned above. The evacuation time, at 375°C, was rather long.

Liebowitz et al. present a considerable body of data which are

* An additional uptake of hydrogen on nickel films contaminated with preadsorbed oxygen has actually been demonstrated by V. Ponec and Z. Knor, Actes 2e congr. intern. catalyse, Paris 1960 2, 2345 (1961).

interpreted to support their view concerning the slow process. Our principal concern in this book is less the rate-determining process and more the actual mechanism of bond formation. Unfortunately, there is not sufficient overlap for us to draw any conclusions concerning this matter. The physical significance of an adsorption site being created by the presence of hydrogen adjacent to the nickel surface is not clear, nor is the decay process developed in detail sufficient for us to relate what is thought to occur to any other observation on which reliance may be placed. But we cannot dismiss the Taylor-Thon mechanism without mention of the extraordinary way in which attention is concentrated on the almost trifling amount of hydrogen actually taken up by the slow process. In their Run 29, for instance, Liebowitz et al. started observations after the (rapid) adsorption of 11.49 ml H_2, and continued until the total sorption was 12.91 ml. This procedure seems .to imply that the rapid process is somehow fundamentally different from the slow process, that the rapid process is somehow inscrutable, and that the progress of surface chemistry is best served by ignoring it.

We shall now present such magnetic data as are available[4,27] on the slow process and will attempt to interpret these data with the aid of the principles already presented in this and earlier chapters. Figure 48 shows magnetization-volume isotherms at 30°C for two commercial nickel-kieselguhr catalyst samples. The isotherms were continued beyond the rapid hydrogen uptake for several days until the processes taking place appeared to be virtually complete. (A time scale is shown, but it will be clear that the time consumed during the rapid process is that required for manipulation and for attainment of thermal equilibrium after each hydrogen increment.) It will be noted that in both cases the "slow"-process hydrogen is a moderate fraction of the whole, that the slope of the isotherm changes for one sample, but that it does not change for the other.

The change in rate of adsorption is so abrupt that little difficulty is experienced in determining, in each case, the volume of hydrogen taken up rapidly or slowly, as the case may be. Those quantities related to the rapid process are designated by the subscript 1, and those related to the slow process by the subscript 2.

Let us assume that the rapid sorption of hydrogen takes place on nickel particles of volume v_1, and the slow on those of volume v_2. Then, if the particles are spheres we may relate the volume of

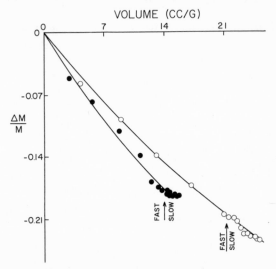

Fig. 48. Magnetization-volume isotherms at room temperature for hydrogen on (●) Universal Oil Products nickel-kieselguhr and (○) Harshaw nickel hydrogenation catalyst. The isotherms have been extended to include the slow process occurring over a period of 72 hours in each case.

the spheres to the hydrogen taken up by each category, as follows:

$$\frac{(N_H)_1}{(N_H)_2} = \frac{n_1 v_1^{2/3}}{n_2 v_2^{2/3}}$$

whence:

$$\frac{v_1}{v_2} = \frac{(N_H)_1^{3/2}}{(N_H)_2^{3/2}} \left(\frac{n_2}{n_1}\right)^{3/2} \tag{7.3}$$

where $(N_H)_1$, $(N_H)_2$ are the total numbers of hydrogen atoms sorbed on all particles of volumes v_1, v_2, respectively, and n_1, n_2, as before, are the total numbers of particles of volumes v_1, v_2, respectively.

Before we proceed with this problem it will be appropriate to say something in support of the assumption that all the particles of nickel are spheres. The evidence for this is tenuous, but ferromagnetic resonance experiments by Hollis and Selwood[28] tend to favor spherical particles, as already indicated (p. 49) for cobalt. Furthermore, such electron microscope evidence as is available tends to favor clusters of spheres for silica-supported nickel.[29-31]

Returning now to Eq. (6.20), $\Delta M_1/\Delta M_2 = (n_H)_1 n_1 v_1/(n_H)_2 n_2 v_2$, we may substitute for v_1/v_2, from Eq. (7.3), and obtain:

$$\frac{\Delta M_1}{\Delta M_2} = \frac{(N_H)_1^{5/2}}{(N_H)_2^{5/2}} \cdot \left(\frac{n_2}{n_1}\right)^{3/2} \tag{7.4}$$

This permits us to find n_1/n_2 from the experimentally determinable quantities $\Delta M_1/\Delta M_2$ and $(N_H)_1/(N_H)_2$, as shown in Fig. 48. The calculations for v_1/v_2, V_1/V_2, and M_1/M_2 are then all readily completed. The experimental data and the derived ratios are shown in Table IX.

TABLE IX

DATA (FROM FIG. 48) AND DERIVED QUANTITIES RELATIVE TO THE "SLOW" ADSORPTION OF HYDROGEN ON NICKEL

Quantity[a]	UOP	HAR
$\Delta M_1/\Delta M_2$ (obs)	30	5.6
$(N_H)_1/(N_H)_2$ (obs)	12.4	8.2
$n_1/n_2 = (\Delta M_2/\Delta M_1)^{2/3}[(N_H)_1/(N_H)_2]^{5/3}$	7.05	39.4
$v_1/v_2 = [(N_H)_1/(N_H)_2]^{3/2}(n_2/n_1)^{3/2}$	2.34	0.597
$V_1/V_2 = v_1 n_1/v_2 n_2$	16.5	23.5
$M_1/M_2 = v_1^2 n_1/v_2^2 n_2$	38.6	14.0

[a] ΔM_1, ΔM_2 are the changes of magnetization produced by rapid and slow hydrogen adsorption, respectively. $(N_H)_1$, $(N_H)_2$ are the total numbers of hydrogen atoms adsorbed rapidly and slowly respectively on particles of individual volumes v_1, v_2, of total volume V_1, V_2, and of magnetizations M_1, M_2.

Table IX shows that the number (n_1) of particles adsorbing hydrogen rapidly is greater than the number (n_2) adsorbing slowly, although the ratio is different in the two samples investigated. The volume (v_1) of the individual particles adsorbing rapidly is a little larger in one sample, a little smaller in the other, than the volume (v_2) of those adsorbing slowly. The total volume (V_1) of "rapid" adsorbent is greater in both samples, and the "slow" adsorbent contributes only moderately to the total magnetization in each case. We see, therefore, that a self-consistent set of data emerges from treatment of the "slow" sorption of hydrogen as in no way fundamentally different from the rapid sorption. Equation (6.20) was derived on the assumption that ϵ was the same for slow and rapid processes. Certainly no major change in ϵ could occur and still have M_1/M_2 and the other ratios turn out in such a reasonable fashion. We may, therefore, conclude that no very obvious change of bond type occurs in going from the one process to the other. It is probable that during the preparation of the samples some nickel nitrate solution becomes trapped in pores and voids in the silica support structure and that this material ultimately results in particles of nickel somewhat less accessible to hydrogen than others.

A related phenomenon is that in which part, but not all, of the nickel may be removed from a nickel-silica preparation by the action of carbon monoxide in forming nickel carbonyl. It is known[4] that this treatment leads to removal of the larger particles, at least so far as the UOP catalyst system is concerned.* The reason for this is quite clear from Table IX—the larger particles of nickel in the UOP catalyst sample are those which are most readily accessible to hydrogen and, presumably, to carbon monoxide.

It is not the purpose of this section to explain the underlying reason why adsorption on some nickel particles should be rapid and on others slow. There is evidence from several sources that the Schuit and De Boer concept of superficial oxidation is correct, or at least partly correct, in some cases. Thus a nickel-silica sample evacuated, after reduction, at 400°C for 60 hours showed a substantial sorption of hydrogen but very little change of magnetization.[27] Such an effect is comprehensible in terms of replacement of adsorbed oxygen by adsorbed hydrogen. We should, however, like to call attention to the study by Doerner[23] who investigated† the sorption of hydrogen on nickel-kieselguhr samples from 0° to 300°C over a pressure range up to 30 atm. Doerner observed the "slow" sorption of hydrogen and he presents convincing evidence that the rate-determining process is diffusion through the micropores, and that this diffusion is probably of the Knudsen type.

We may now add to our summary of conclusions (p. 117) concerning the chemisorption of hydrogen on silica-supported nickel. The so-called activated adsorption of hydrogen is probably a diffusion-controlled process on to less accessible particles of nickel. (In some cases the presence of adsorbed oxygen probably

---

* This observation has recently been confirmed by Yates and Garland[32] on the basis of infrared absorption spectroscopy.

† To the best of the author's knowledge this thesis has never been published in one of the scientific journals. This is regrettable.

contributes to the slow process.) A particle of nickel difficult of access to hydrogen must be more so to carbon monoxide, and to benzene; it must be very difficult for a molecule of nickel carbonyl to escape from such a particle. Under the circumstances it is difficult to see how the slow process can be of much significance in heterogeneous catalysis.

4. The Effect of Deviations from Collective Paramagnetism

It will be recalled that all the preceding argument applies only to those assemblies of small particles in which collective paramagnetism is exhibited. The several criteria for collective paramagnetism have already been presented (p. 94). In this section we shall refer to some of the (rather bizarre) effects observed when gases are adsorbed on samples which, for one reason or another, exhibit deviations from collective paramagnetism.

If a typical nickel-kieselguhr sample is sintered, after reduction, at about 600°C for several hours, the nickel particles increase in size,[33] and no longer show the characteristic behavior associated with collective paramagnetism. If now hydrogen is permitted to adsorb on the nickel surface the changes of magnetization observed may be quite different from those described in preceding sections. A good example of these effects is shown in Fig. 49, obtained by

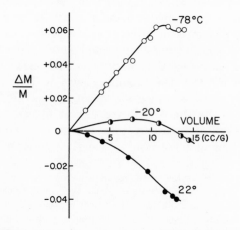

FIG. 49. Magnetization-volume isotherms for hydrogen on nickel. The nickel particle size is a little too large to exhibit collective paramagnetism as the temperature is lowered (after Leak).

Leak[34] in which the change of magnetization is positive* at $-78°C$; at first positive but later negative with increasing surface coverage at $-20°$; and negative, though not linear, at $22°$.

The conditions leading to a positive slope for the magnetization-volume isotherm in the nickel-hydrogen, and certain other, systems appear to be as follows:

(1) The nickel particles should be rather larger than those discussed at length above. A diameter of about 100 A seems to be effective.

(2) Low temperatures are favorable.

(3) Low fields may produce the effect, although under certain conditions positive slopes are obtained with moderately high fields.

(4) The use of AC fields favors the effect although here again, certain conditions may lead to positive slopes in DC fields.

The reason for these anomalies seems almost certainly to be that suggested by Dietz,[18] namely, that a particle which is, through size, shape anisotropy, or for some other reason unable to turn its magnetic axis freely in the applied field will, of course, not exhibit its normal magnetization. This effect will be aggravated in larger particles, those of elongated shape, at low temperatures, and by the use of AC fields. It seems probable that the magnetic anisotropy of such particles is actually decreased by the presence of adsorbed hydrogen. A related effect, referred to as "spheroidization," in elongated iron-cobalt single domain particles embedded in a thermosetting plastic is described by Mendelsohn and Norman.[35] The effect of decreasing anisotropy could be to make rotation in the applied field easier and this, in turn, would make the apparent magnetization greater. The example shown in Fig. 49 at $-20°C$ is an interesting case in which the anisotropy having been at least partly destroyed by the first increments of hydrogen, the normal "negative" effect of hydrogen then becomes apparent. It seems probable, in view of these results, that the "positive" effects have

---

* The positive effect, which had previously been observed,[4] was at first erroneously attributed to a change of bond type, and thought to indicate formation of negative hydride ions, H^-, which by removal of electrons from the d band of the nickel were thought to result in more unpaired electrons in the same fashion that oxidation of Fe^{2+} to Fe^{3+} raises the magnetic moment from about 4.9 Bohr magnetons to 5.9. This interpretation of the positive effect in the nickel-hydrogen system is almost certainly wrong, but the idea that H^- may be present is not necessarily excluded by this conclusion.

a logical explanation and that they do not indicate any change of bond type. Some further reference to this phenomenon will be made later when we come to consider the adsorption of oxygen on nickel.

5. Hydrogen on Cobalt

The peculiarly low value of ϵ, the change of magnetic moment produced by an atom of hydrogen adsorbed on cobalt, has already been described (p. 69). In this section there will be presented some rather fragmentary results of permeameter studies made for the purpose of obtaining magnetization-volume isotherms on supported cobalt systems. As previously pointed out (p. 40), the structure of cobalt metal is such that collective paramagnetism is not observed until the particle size becomes much smaller than is the case for nickel. This means, in practice, that correspondingly higher temperatures must be used for the cobalt studies [because of the temperature term in Eq. (3.8)]. Fortunately, the very high Curie temperature of cobalt makes such studies possible, although the phase transition occurring in the neighborhood of 400°C may cause some difficulties at low fields.[36]

Figure 50 shows the fractional change of permeameter emf for

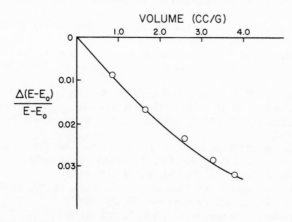

FIG. 50. A plot of secondary emf versus hydrogen adsorbed on cobalt-silica at 450°C (after Abeledo).

hydrogen adsorbed on silica-supported cobalt at 355°C.[19] The general trend of results seems to be similar to that observed for hydrogen on nickel at more moderate temperatures. The results obtained to date are significant in that they suggest a method for studying chemisorption at temperatures considerably more elevated than those possible with nickel.

Studies related to these have recently been reported by Romanowski[37] who measured the magnetizations of reduced cobalt supported on charcoal, gamma-alumina, and magnesia. The degree of reduction was, in most cases, far from complete, but an influence of adsorbed hydrogen in changing the (extrapolated) saturation magnetization was definitely observed. The magnetic measurements were made at 290°K and evidence is presented which suggests some "electronic" influence of semiconducting cobaltous oxide on the saturation magnetization of cobalt metal. But the data would appear to lend themselves equally well to the "anisotropy" explanation given above.

REFERENCES

1. P. W. Selwood, *Rev. inst. franç. pétrole et Ann. combustibles liquides* **13**, 1656 (1958).
2. I. Den Besten and P. W. Selwood, *J. Phys. Chem.* **66**, 450 (1962).
3. P. W. Selwood, *J. Am. Chem. Soc.* **78**, 3893 (1956).
4. E. L. Lee, J. A. Sabatka, and P. W. Selwood, *J. Am. Chem. Soc.* **79**, 5391 (1957).
5. J. J. Broeder, L. L. van Reijen, W. M. H. Sachtler, and G. C. A. Schuit, *Z. Elektrochem.* **60**, 838 (1956).
6. J. A. Silvent and P. W. Selwood, *J. Am. Chem. Soc.* **83**, 1034 (1961).
7. B. M. W. Trapnell, "Chemisorption," p. 10. Academic Press, New York, 1955.
8. H. Sadek and H. S. Taylor, *J. Am. Chem. Soc.* **72**, 1168 (1950).
9. G. C. A. Schuit, N. H. de Boer, G. J. H. Dorgelo, and L. L. van Reijen, *in* "Chemisorption" (W. E. Garner, ed.), p. 44. Academic Press, New York, 1957.
10. L. Vaska and P. W. Selwood, *J. Am. Chem. Soc.* **80**, 1331 (1958).
11. I. Den Besten and P. W. Selwood, *J. Catalysis* **1**, 93 (1962).
12. S. Umeda, *J. Japan Soc. Powder Met.* **8**, 159 (1961).
13. P. M. Gundry and F. C. Tompkins, *Quart. Revs.* **14**, 269 (1960).
14. R. Gomer, *Discussions Faraday Soc.* **28**, 23 (1959).
15. R. Suhrmann, *Advances in Catalysis* **7**, 303 (1955).
16. W. M. H. Sachtler, *J. Chem. Phys.* **25**, 751 (1956).
17. W. M. H. Sachtler and G. J. H. Dorgelo, *Bull. soc. chim. Belges* **67**, 465 (1958).
18. R. E. Dietz, Thesis, Northwestern University, Evanston, Illinois, 1960.
19. C. R. Abeledo, Doctoral Dissertation, Northwestern University, Evanston, Illinois, 1961.

20. T. B. Grimley, *J. Phys. and Chem. Solids* **14**, 227 (1960).
21. "Metals Reference Book" (C. J. Smithells, ed.) 2nd ed., Vol. II, pp. 533–45. Butterworths, London, 1955.
22. H. S. Taylor, *J. Am. Chem. Soc.* **53**, 578 (1931).
23. W. A. Doerner, Dissertation, University of Michigan, Ann Arbor, Michigan, 1952.
24. G. C. A. Schuit and N. H. De Boer, *Rec. trav. chim.* **72**, 909 (1953).
25. L. Liebowitz, M. J. D. Low, and H. A. Taylor, *J. Phys. Chem.* **62**, 471 (1958).
26. H. A. Taylor and N. Thon, *J. Am. Chem. Soc.* **74**, 4169 (1952).
27. P. W. Selwood, to be published.
28. D. P. Hollis and P. W. Selwood, *J. Chem. Phys.* **35**, 378 (1961).
29. G. C. A. Schuit and L. L. van Reijen, *Advances in Catalysis* **10**, 259 (1958).
30. H. C. Corbet, *in* "Fourth International Congress on Electron Microscopy, Berlin 1958," p. 47. Springer, Berlin, 1960.
31. S. Shishido, *Nippon Kagaku Zasshi* **81**, 679 (1960).
32. J. T. Yates and C. W. Garland, *J. Phys. Chem.* **65**, 617 (1961).
33. W. Heukelom, J. J. Broeder, and L. L. van Reijen, *J. chim. phys.* **51**, 474 (1954).
34. R. J. Leak and P. W. Selwood, *J. Phys. Chem.* **64**, 1114 (1960).
35. L. I. Mendelsohn and R. S. Norman, *J. Appl. Phys.* **30**, 142S (1959).
36. R. M. Bozorth, "Ferromagnetism," p. 265. Van Nostrand, New York, 1951.
37. W. Romanowski, *Chemia Stosowana* **2**, 225 (1961).

CHAPTER VIII

Hydrogen Sulfide, Cyclohexane, Cyclohexene, and Benzene

The last three chapters in this book will be devoted to the magnetic method for determining the number of bonds formed on a metal adsorbent by almost any adsorbate. In this chapter the method itself will first be presented, after which some notably successful applications will be described. Our obvious purpose here is to consolidate our position before proceeding to more controversial areas in later chapters.

1. Theory of the Method for Determining Bond Number

The possibility[1] that magnetic measurements may be used to find the number of chemisorptive bonds has been developed[2] in some detail. It has proved to be the most useful feature of the magnetic approach—or at least this is true so far as heterogeneous catalysis is concerned. We shall, therefore, devote considerable attention to the method which consists, in brief, of comparing the slope of the magnetization-volume isotherm with that of hydrogen, under conditions as nearly alike as possible. While this presentation will be limited chiefly to nickel as adsorbent, the adsorbate may be any molecule, with the possible exception of one such as water which is strongly adsorbed on the silica support.

We shall assume that hydrogen is chemisorbed on nickel over the room temperature range, and higher, by a dissociative mechanism leading to Ni—H bonds. The evidence for this was presented in detail in the preceding chapter. It is true that we cannot yet define the kind of surface bond formed, and we cannot definitely say that the hydrogen atom is attached to one, and only one, atom of nickel. But these considerations do not concern us in

130

the present section. The essential points here are that each hydrogen molecule should form two bonds to nickel, that the nature of the nickel-hydrogen bond should not change with increasing surface coverage, and that, within wide limits, it should not change with temperature.

It would, of course, be possible to measure the number of electron spins paired per molecule of adsorbate by direct measurement of the saturation magnetization as described for hydrogen in Chapter IV, but the tedious processes of carrying out the adsorptions at room temperature but measuring the magnetizations at 4.2°K do not seem attractive. The low-frequency AC permeameter offers, on the other hand, a wealth of convenience which is by no means all counteracted by some lack of definiteness in the interpretation. All the data described in this and the following chapters were obtained on the permeameter. Examination of Eq. (6.6), or better (6.15) shows that $\Delta M/M$ is, indeed, proportional to $\Delta M_s/M_s$ provided that collective paramagnetism is exhibited. But it is obvious that comparison of isotherms produced by one adsorbate and another is a valid procedure only if the adsorbent remains in the same condition with respect to volume (and saturation magnetization), temperature, and distribution of particle sizes. These conditions may be met approximately if the identical sample is completely reduced and evacuated, used to obtain a hydrogen isotherm at the desired temperature, then evacuated again with no significant nickel particle growth prior to obtaining of data for the second isotherm. Actually, these conditions offer no more difficulty than many standard procedures in physical chemistry.

Having obtained our two isotherms we may relate the slopes to the number of bonds formed thus, if 1 cc of vapor X_2 lowers the magnetization just twice as much as 1 cc of H_2, then molecule X_2 is held to the nickel by four bonds.

This procedure for obtaining the number of bonds is based on several assumptions. First, may we say that $\Delta M/M$ varies directly as the number of bonds formed? It is certainly possible from Eq. (6.15) and its agreement with experiment to say that $\Delta M/M$ varies directly with the number of electron spins paired per molecule adsorbed.[*] Then our question actually becomes: Is it possible to differentiate between a kind of double bond Ni=X and

---

[*] Subject, of course, to the possible spin-orbital complication discussed on p. 81.

two single bonds to two different but adjacent atoms thus: Ni—X—Ni? No final answer to this question seems yet possible. We have, however, presented more than a little evidence pointing to localization of the Ni—H bond. If localization actually occurs then Ni=X might be expected to have half the magnetization effect of Ni—X—Ni. We shall be forced to leave this question unanswered for the time being. Studies on the cobalt-hydrogen system,[3] which were expected to throw light on this problem, have not yet done so.

Another assumption is that $\Delta M/M$ for any bond Ni—X is the same as for Ni—H. Here again we must rely mostly on internal evidence. If the moment of one nickel atom is destroyed by its union with one hydrogen atom (as is approximately the case), then it seems reasonable that Ni—X and Ni=X would have the same effect. Unfortunately, less than one-third of the moment of cobalt is destroyed by one atom of hydrogen, and we shall see later that adsorbed atoms such as sulfur and carbon seem able to destroy the moments of two and three nickel atoms, respectively.

The above remarks are intended to suggest that what we mean by "bond number" is actually "number of adsorbent atoms bonded." We shall find considerable internal consistency in the results obtained with a variety of adsorbates. Most important, however, is the fact that there are available chemical methods for investigating these problems. It is certainly to be hoped that the magnetic method will yield results in agreement with some of these chemical methods and, fortunately, this is actually the case.

2. Hydrogen Sulfide and Dimethyl Sulfide

Figure 51 shows magnetization-volume isotherms obtained by Den Besten[4] for hydrogen and hydrogen sulfide on a commercial nickel-kieselguhr (UOP) containing 52.8% of nickel reduced and evacuated in the usual way. The temperature of adsorption and of measurement was 25°C. This pair of isotherms, typical of many, shows that the slope of the hydrogen sulfide isotherm is, within ±5%, twice that of hydrogen. The implication is that hydrogen sulfide is, under these conditions, dissociatively adsorbed thus:

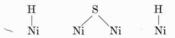

(there being, of course, no implication as to whether hydrogen or

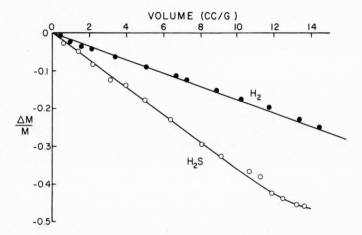

FIG. 51. Magnetization-volume adsorption isotherms for hydrogen sulfide at room temperature on nickel-kieselguhr (after Den Besten).

sulfur is attached on or between the nickel atoms). Isotherms obtained at 0° and at 115°C give almost identical results. We may, therefore, conclude on the basis of the magnetic data that dissociation of the hydrogen sulfide is complete, and that there is no change of bond type over the temperature range investigated.

Support for the view stated above is available from at least two sources. First, even up to quite appreciable pressures the only gas found in the free space over the catalyst is hydrogen. Furthermore, provided that surface coverage is moderate it is possible at 360°C to evacuate nearly all the hydrogen as molecular hydrogen, but none as hydrogen sulfide. Thus, after the adsorption of 1.43 cc H_2S per gram Ni at 25°C, one may remove 1.36 cc H_2 (as such) by evacuation at 360°. It is difficult to see how these results could be obtained if complete dissociative adsorption had not occurred.

The most convincing agreement is, however, that provided by Kemball[5] who shows that, in the temperature region covered, two hydrogen atoms, per molecule of hydrogen sulfide adsorbed, are readily exchanged for deuterium.

At higher surface coverages of hydrogen sulfide, more complicated effects occur. One of these is that as the pressure over the sample becomes appreciable, it may be observed that each additional increment of adsorbed hydrogen sulfide results in the rather slow liberation of hydrogen, so that some increase of hydrogen pressure actually occurs. This is in sharp contrast to the slow disappearance of hydrogen from the gas phase when, under similar conditions, hydrogen itself is the adsorbate. This effect is probably due to progressive dissociation of chemisorbed hydrogen sulfide molecules on a surface which, because it is already nearly covered with hydrogen, can accept no more hydrogen. But the more tightly bound sulfur is, nevertheless, readily accepted.

It is also found that after surface coverage is virtually complete, it becomes impossible to remove more than about 70% of the hydrogen either by evacuation at 360°C or by exchange with deuterium. The reason for this is not clear but it may be related to blocking of the surface by sulfur, which retards egress of interstitial hydrogen. In this connection it is found that addition of hydrogen sulfide to a surface already covered by hydrogen results in liberation of some hydrogen from the surface. It will also be noted that the cumulative loss of magnetization caused by hydrogen sulfide is nearly twice that produced by hydrogen as surface coverage appears to be nearing completion. If our views concerning the ability of hydrogen to destroy the magnetization of each surface nickel atom are correct, then it is obvious that the sulfur from hydrogen sulfide must involve more than the surface layer of nickel atoms. This view is also consistent with Kemball's findings,[5] the only important difference being that changes occur somewhat more readily and at lower temperatures on evaporated films. When we investigate oxygen adsorption (in a later chapter) we shall find that true chemisorption, as opposed to multilayer oxide formation, is the exception rather than the rule.

Dimethyl sulfide as a catalyst poison is even more notorious than hydrogen sulfide. Magnetization-volume isotherms for this adsorbate have been obtained by Den Besten,[4] as shown in Fig. 52 for the temperatures 25° and 120°C. The initial slope at room temperature is only slightly greater than that for hydrogen. This suggests bonding to two nickel atoms as follows:

$$
\begin{array}{ccc}
CH_3 & & CH_3 \\
\diagdown & & \diagup \\
& S & \\
\diagup & & \diagdown \\
Ni & & Ni
\end{array}
$$

in a manner resembling that suggested by Kemball[5] for hydrogen sulfide on nickel films in the neighborhood of −80°C.

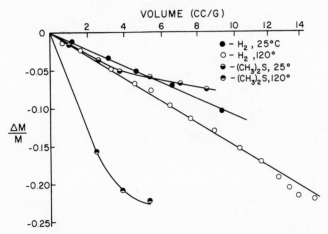

FIG. 52. Magnetization-volume isotherms for hydrogen and for dimethyl sulfide on nickel-kieselguhr at 25° and at 120°C (after Den Besten).

The adsorption mechanism of dimethyl sulfide is quite sensitive to temperature. At 120°C the magnetic data indicate quite extensive dissociation—the initial slope being consistent with at least 10 bonds being formed per molecule adsorbed. This conclusion is confirmed by the nature of the desorption products obtained as the temperature is gradually raised to 400°C. These products include methane, ethane, and hydrogen, but no sulfur compound. Our view of catalyst poisoning by Lewis bases such as hydrogen sulfide and dimethyl sulfide is, therefore, the classical one of atoms, such as sulfur, bonded to the catalyst sites more strongly than hydrogen, or at least more strongly than adsorbed hydrogen in a reactive state. The greater efficiency of dimethyl sulfide as a catalyst poison appears to be related to its greater covering power per molecule adsorbed.

There appear to be no previous studies with which to compare the magnetic data on dimethyl sulfide, excepting that mentioned on p. 16.

3. Cyclohexane[6]

Figure 53 shows magnetization-volume isotherms for cyclohexane on nickel-kieselguhr at 25° and at 150°C. The isotherm at room temperature is not easy to interpret accurately because the large van der Waals adsorption tends to obscure the true chemisorption,

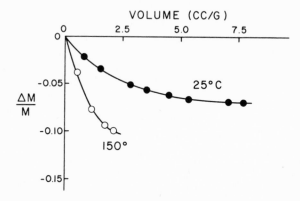

FIG. 53. Magnetization-volume isotherms for cyclohexane on nickel-kieselguhr at 25° and at 150°C.

although it is obvious that some chemisorption occurs. But at 150° the van der Waals adsorption is much diminished, and the only complication is some slight evidence of progressive dissociation which causes a pressure rise of a few millimeters per hour.

The effect on the magnetization of a molecule of cyclohexane at 150° is four to six times that of a molecule of hydrogen. (The total volume adsorbable is, of course, much diminished as compared with hydrogen.) We may, therefore, say that under these conditions cyclohexane is held to the nickel by a minimum of eight bonds. It is clear that a saturated hydrocarbon could be chemisorbed only through a dissociative mechanism. It appears that cyclohexane is adsorbed on nickel by dissociating four hydrogen atoms, thus:

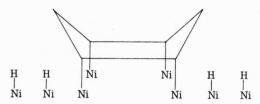

although it is probable that some further dissociation occurs at 150° and above.

This view concerning the chemisorption of cyclohexane receives strong support from the deuterium exchange results of Galwey and Kemball.[7] These authors carried out their experiments from about

—20° to 210°C, plotting the number of hydrogen atoms initially exchangeable for deuterium per molecule of cyclohexane adsorbed on reduced nickel-silica. (The adsorbent was quite similar to that used in the magnetic work.) The number of hydrogen atoms so exchangeable rises from zero at —20°C to four at 30°C, and remains constant at four to 200°C. Agreement with conclusions based on the magnetic data is, therefore, very satisfactory.

4. Cyclohexene

Figure 54 shows magnetization-volume isotherms obtained for

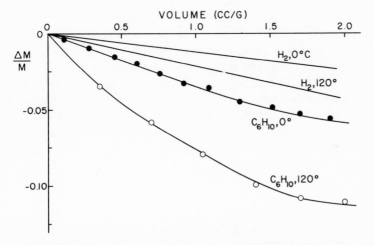

FIG. 54. Magnetization-volume isotherms for hydrogen and for cyclohexene on nickel-kieselguhr at 0° and at 120°C (after Den Besten).

cyclohexene by Den Besten[4] at 0° and 120°C. Hydrogen isotherms at the corresponding temperatures are also shown. At 120°C the maximum pressure over the sample was 7 mm. The initial slope of the isotherm obtained at 25°C was the same as that at 0°C, namely, 2.65 times that of hydrogen under the same conditions. At 120°C the initial slope was approximately four times that of hydrogen. These results indicate that in the temperature range 0°–25°C the average molecule of cyclohexene forms about 5.3 bonds with nickel, and that at 120°C as many as eight bonds are formed.

Before proceeding to the interpretation of these data we shall refer to the paper by Galwey and Kemball.[8] The adsorbent was a nickel-silica similar to that used in the magnetic studies. While it is not possible to estimate the extent of surface coverage with cyclohexene it is probable that the coverage was comparable with that shown in Fig. 54 for the first increment or two. The number of hydrogen atoms readily exchangeable for deuterium was then determined in the manner to which reference has already been made on several occasions. The results stated are that two hydrogen atoms per molecule of cyclohexene were exchanged after adsorption at 0°C or higher, and that four more hydrogen atoms were exchanged when the sample was heated in excess deuterium to 100°C. The lower temperature experiment is strictly comparable with the magnetic, the 100° run rather less so. Results were also reported on the exchange observed for cyclohexene adsorbed at 0°C, then heated to various temperatures up to 180°. The number of exchangeable hydrogen atoms remained at three in this case. There are no magnetic data strictly comparable to this last exchange result. It has, however, been found by the writer that a hydrocarbon molecule adsorbed at a lower temperature is somewhat more resistant to (further) dissociation on progressive heating than is a molecule admitted to the adsorbent at a more elevated temperature. The reason for this is doubtless that during the adsorption process considerable excess energy is made available during formation of the Ni—H and Ni—C bonds, and that this excess energy may accelerate dissociative processes. The precision claimed by Galwey and Kemball is ±0.5 $\tilde{n}$, where $\tilde{n}$ is the number of H atoms exchanged per molecule adsorbed. The precision of the magnetic data is believed to be considerably better than this and perhaps the equivalent of ±0.1 $\tilde{n}$.

The mechanism of adsorption suggested by Galwey and Kemball for the room temperature case is dissociative, yielding two hydrogen atoms and a C_6H_8 group. These may be adsorbed as follows:

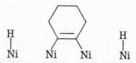

although precisely how the C_6H_8 radical is adsorbed is not completely established by any of these results. In any event, the concept of two hydrogen atoms being dissociated is, within probable experimental error, in agreement with the magnetic data, especially if we recall that complete fragmentation of a small fraction of the cyclohexene molecules could have a marked effect on the slope of the magnetization isotherm. The apparent total number of bonds formed per adsorbed molecule is 5.3 ±0.5 (magnetic) and 4.0 ±1.0 (exchange). There is no reason to believe that this mechanism of adsorption is related to the well-known disproportionation of cyclohexene over nickel-silica.[9] The disproportionation occurs at temperatures above those at which the number of dissociated hydrogen atoms is limited to two.

Another kind of evidence offers further support for the views expressed above. While adsorbed cyclohexene may be hydrogenated at moderately elevated temperatures, it is definitely not hydrogenated at −78°C. In this experiment a pressure-volume isotherm for hydrogen is obtained at −78°. The sample is then heated and evacuated to remove all the hydrogen, after which a measured volume of cyclohexene is admitted at room temperature. The sample is now cooled to −78° again, and a second pressure-volume isotherm for hydrogen is obtained (see Fig. 55). In a typical run

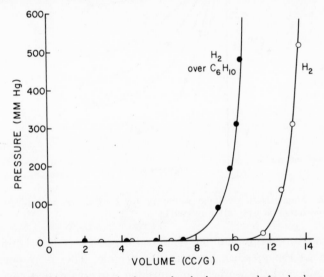

Fig. 55. Pressure-volume isotherms for hydrogen and for hydrogen over preadsorbed cyclohexene on nickel-kieselguhr at −78°C (after Den Besten).

it was found[4] that adsorption of 1.50 cc (STP) of cyclohexene vapor diminished the volume of hydrogen adsorbed (at 1 atm) at −78° by 3.10 cc, all volumes being given per gram of nickel.

These results show that one molecule of cyclohexene chemisorbed at room temperature is able to deny access to the surface of two molecules of hydrogen—the hydrogen being, of course, admitted under conditions which prohibit the hydrogenation reaction. This result is in complete agreement with the views expressed above concerning the mechanism of cyclohexene adsorption.* We shall have further reference to this useful technique later.

5. Benzene[6,10]

Magnetization-volume isotherms for benzene on nickel-silica over a wide range of adsorption temperatures are shown in Fig. 56.

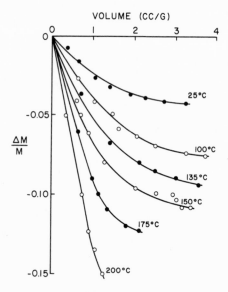

FIG. 56. Magnetization-volume isotherms for benzene on nickel-kieselguhr at various temperatures (after Silvent).

At room temperature van der Waals adsorption becomes significant as soon as the pressure becomes appreciable. This is responsible for the slope of the isotherm diminishing with increasing coverage at the lower temperatures. It will be noted also that the total volume, measured as vapor corrected to standard conditions, is

* It was earlier pointed out (p. 112) that the presence of the preadsorbed cyclohexene does not alter the slope of the magnetization-volume isotherm for hydrogen.

much smaller for benzene than for hydrogen. The first increments of benzene show strong thermal transients like those shown by hydrogen (p. 101). Furthermore, at the higher temperatures progressive dissociation occurs and, at 200°C these effects tend to complicate interpretation of the data.

In spite of these several difficulties it is quite clear that benzene enters into electronic interaction with nickel. This is to say that benzene is chemisorbed over the whole temperature range represented in Fig. 56. It is also clear that the temperature of adsorption has a marked effect on the number of bonds formed per molecule of adsorbed benzene. Table X shows the average number of bonds formed, as a function of temperature. (These data are, of course, merely twice the ratio of isotherm slopes for benzene and hydrogen respectively at the several temperatures.) These results may be

TABLE X
AVERAGE BOND NUMBER PER ADSORBED BENZENE MOLECULE

Adsorption temperature (°C)	25	100	150	200
Bonds formed	5.5	6.0	10.5	18

interpreted as showing associative adsorption up to about 130°C, above which dissociative adsorption becomes increasingly important. Six bonds formed by benzene suggest that the molecule lies flat; more than six bonds suggest not only dissociation of hydrogen but some degree of carbon-carbon bond rupture.

Some evidence that these views are correct may be obtained from two experiments conducted parallel to the magnetic studies. The first is the rather obvious one of attempting to hydrogenate the adsorbed molecules or molecular fragments. It is found that if hydrogen is allowed to flow over preadsorbed benzene at 100°C, it is possible to recover, by freezing, most of the original adsorbate in the form of cyclohexane. But if the adsorption step is conducted at 200°, the effluent vapor contains an appreciable fraction of lower molecular weight hydrocarbons, although a substantial amount of cyclohexane still appears. This result suggests that the very large number of bonds indicated by the magnetic data for the higher adsorption temperatures is made up of hydrogen-nickel bonds plus carbon-nickel bonds, but with the carbon still mostly present in the form of six-membered rings.

Additional evidence supporting these views is obtained from the second experiment, which is to find the volume of hydrogen denied access to the nickel (at $-78°C$) by the presence of a measured volume of preadsorbed benzene, as already described for cyclohexene (p. 139). The result of this experiment is that at a benzene adsorption temperature of 25°C one molecule of benzene denies access to the nickel of 2.6 molecules of hydrogen. This number rises with increasing benzene adsorption temperature until at 200° it is 8.6.

Our over-all conclusions respecting benzene on nickel are, therefore, that at room temperature the molecule is adsorbed flat without dissociation, but that above about 130°C extensive dissociation of hydrogen plus some carbon-carbon bond rupture occur. There is some evidence also that progressive *in situ* dissociation may be observed magnetically at appropriate temperatures.[10] It might be hoped that the wealth of literature on benzene hydrogenation over nickel would yield information with which we might compare the conclusions based on magnetic data. Such, unfortunately, is not the case. The loss of aromatic character for chemisorbed benzene on nickel is apparently shown by infrared absorption spectroscopy,[11] and the changes of electrical conductivity of nickel films after benzene adsorption have been interpreted[12] as showing the presence of six bonds (for the room temperature case). Pitkethly and Goble[13] have presented kinetic evidence for flat chemisorption of benzene on platinum.

6. The Interaction of Benzene and Hydrogen on Nickel

Useful speculations concerning the mechanism by which benzene is hydrogenated over a metal are few in number. Balandin[14] proposed many years ago that the benzene molecule must lie flat, being adsorbed by six metal-carbon bonds. Balandin observed that this kind of adsorption could most readily occur on metal crystal faces having an appropriate geometric spacing of atoms. This idea has been much discussed and developed by its author and by others; there is considerable attractiveness to the proposal.

It seems to the present writer that while benzene may certainly be chemisorbed on nickel, and that the magnetic data point fairly clearly at flat, six-point adsorption, yet there is no demonstrated requirement that such adsorption is necessary for hydrogenation. Hydrogenation must of necessity take place in the presence of hydrogen.* The hydrogen must then be chemisorbed and, under normal hydrogenation conditions, must provide a blanket preventing access of benzene to the metal surface. The benzene comes in contact not with nickel but rather with hydrogen. Thanks to the nickel, the hydrogen is in a peculiarly available and reactive state, with Ni—H bond strengths attenuated to almost any necessary degree. It is certainly possible that a geometric factor is operative, but if this is so, the factor functions through favorable geometric arrangement of the hydrogen which is, of course, held in a pattern determined primarily by the nickel. The incoming benzene molecule may then strike the appropriate pattern of six hydrogen atoms and depart as cyclohexane, leaving six vacant sites to be filled virtually instantaneously by fresh hydrogen.

There is nothing new in the idea expressed above. Let us see what support, if any, it receives from the magnetic studies.[6] If a nickel surface already partly covered with benzene is then treated with hydrogen it will be found, as shown in Fig. 57, that the hydro-

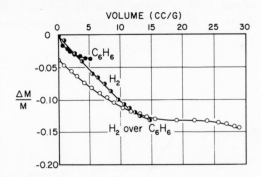

FIG. 57. Magnetization-volume isotherms for hydrogen, for benzene, and for hydrogen over preadsorbed benzene on nickel-kieselguhr, all at room temperature.

gen is preferentially chemisorbed rather than being used for hydrogenation. This is proved because the hydrogen isotherm slope is the same as if no benzene had been present.

Now if more hydrogen is added it will be found (Fig. 57) that the isotherm becomes more nearly horizontal, showing that no net change of bond number is occurring. This is consistent with the

* Excluding, of course, the more complicated reactions of self-hydrogenation, such as occur readily with ethylene and with cyclohexene.

view that hydrogen is now being used for hydrogenation, that nickel-carbon bonds are being broken, and that the vacant nickel sites are being replaced by nickel-hydrogen bonds. The total volume of hydrogen taken up in this process is just that required to hydrogenate all the benzene, plus enough to cover the nickel. The implication is then that while benzene may readily be hydrogenated at room temperature, yet the benzene cannot be hydrogenated directly without prior desorption. When the bond strengths have been weakened to an appropriate degree by increasing surface coverage, then, and then only, can the benzene be desorbed. As soon as the benzene is desorbed from the nickel it is free to pick up six adsorbed hydrogen atoms. The strong van der Waals adsorption probably prevents any emergence of cyclohexane in the gas phase. If preadsorbed benzene cannot be hydrogenated at −77°C, it is probably because the hydrogen is unable to displace the benzene at this temperature.

Further evidence is obtained by covering the nickel surface, at room temperature, with a partial layer of hydrogen, then adding benzene as shown in Fig. 58. Under these circumstances the ben-

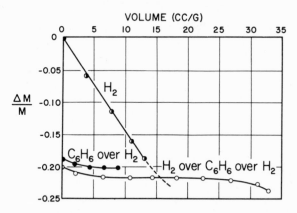

Fig. 58. Magnetization-volume isotherms for hydrogen, for benzene over preadsorbed hydrogen, and for hydrogen over benzene over preadsorbed hydrogen on nickel-kieselguhr, all at room temperature.

zene causes a negligible change of magnetization. Yet, if any appreciable quantity of preadsorbed hydrogen were being used for hydrogenation, then the subsequent addition of more hydrogen would cause a substantial loss of magnetization. But as shown in

Fig. 58, this loss does not occur. We must, therefore, conclude that in the presence of somewhat tightly bound hydrogen the benzene molecule can neither take hydrogen away from the nickel nor can it find an appropriate site for its own chemisorption. But when additional hydrogen is added, the pressure-volume relationship shows that quantitative hydrogenation readily occurs.

These several considerations make it possible to demonstrate the participation of d electrons during an actual catalytic process. A nickel-silica sample reduced and evacuated as usual is covered to about one atmosphere with hydrogen, then sealed off. A few drops of benzene previously introduced and kept frozen adjacent to the catalyst is then allowed to melt. As the benzene diffuses to the nickel surface, some hydrogenation occurs. This causes diminution of the hydrogen pressure which, in turn, leaves some nickel sites bare. The attendant rise of magnetization is readily observed as hydrogenation proceeds (Fig. 59).

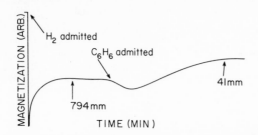

FIG. 59. Magnetization changes occurring in nickel-kieselguhr at room temperature as benzene is admitted to a sample covered with preadsorbed hydrogen.

The several examples presented in this chapter have, it is hoped, presented a strong case for the basic assumptions involved in the method. It seems to the author that the weight of evidence supports the view that a bond Ni—X has the same effect on the magnetization as a bond Ni—H. It seems also to the author that the bond system Ni—X—Ni has twice the effect of Ni—X, and that if such a bond as Ni=X occurs it has the same effect as Ni—X. Our chief assumption here is, of course, that our quantity ϵ (p. 68) is the same for Ni—X as it is for Ni—H. The reader may wonder why we do not measure ϵ for Ni—X directly by saturation magnetization studies; it is certainly possible to do this experiment with almost any molecule as adsorbate. The difficulty is that with the possible exception of hydrogen, there is, in the whole realm of surface chemistry, not one molecule for which the mode of chemisorption is definitely known.

REFERENCES

1. P. W. Selwood, *J. Am. Chem. Soc.* **79**, 3346 (1957).
2. P. W. Selwood, *Actes 2e congr. intern. catalyse, Paris, 1960* **2**, 1795 (1961).
3. C. Abeledo and P. W. Selwood, to be published.
4. I. E. Den Besten and P. W. Selwood, *J. Catalysis* **1**, 93 (1962).
5. C. Kemball, *Actes 2e congr. intern. catalyse, Paris, 1960* **2**, 1811 (1961).
6. P. W. Selwood, *J. Am. Chem. Soc.* **79**, 4637 (1957).
7. A. K. Galwey and C. Kemball, *Trans. Faraday Soc.* **55**, 1959 (1959).
8. A. K. Galwey and C. Kemball, *Actes 2e congr. intern. catalyse, Paris, 1960* **2**, 1063 (1961).
9. B. B. Corson, *in* "Catalysis" (P. H. Emmett, ed.), Vol. III, p. 96. Reinhold, New York, 1955.
10. J. A. Silvent and P. W. Selwood, *J. Am. Chem. Soc.* **83**, 1033 (1961).
11. R. P. Eischens, Personal communication, 1960.
12. R. Suhrmann, G. Kruger, and G. Wedler, *Naturwissenschaften* **47**, 441 (1960).
13. R. C. Pitkethly and A. G. Goble, *Actes 2e congr. intern. catalyse, Paris, 1960* **2**, 1851 (1961).
14. A. A. Balandin, *Z. physik. Chem. (Leipzig)* **B2**, 289 (1929).

CHAPTER IX

Ethane, Ethylene, and Acetylene

No problems in surface chemistry have been more hotly debated than the adsorption and hydrogenation mechanisms for ethylene; and few debates have resulted in such meager conclusions. We shall not be able to resolve many of the controversial questions involved but we shall throw some light on certain aspects of the adsorption *impasse*. Perhaps it will be possible to suggest some routes leading to the final answers.

1. Ethane

It is sometimes stated that ethane is not chemisorbed to any extent on nickel,[1] or that, together with iron and cobalt, nickel is almost completely inactive toward ethane sorption.[2] These statements are quite misleading—Trapnell's[2] own data show a small but quite definite chemisorption, and the exchange data of Kemball and others, summarized by Anderson,[3] show clearly that ethane may undergo exchange over nickel at quite moderate temperatures. It is difficult to see how ethane can exchange any hydrogen for deuterium unless the molecule of ethane is first dissociatively adsorbed. This view that ethane is indeed chemisorbed on nickel receives further confirmation from the infrared absorption spectrum obtained by Eischens and Pliskin[4] who show that on a bare nickel surface at 35°C ethane gives a spectrum similar to that of ethylene on bare nickel.

Figure 60 shows magnetization-volume isotherms at 27°C for ethane on a bare nickel-kieselguhr surface, and also on the same surface almost completely covered by preadsorbed hydrogen.[5] There is an appreciable van der Waals adsorption at this temperature, but the data clearly show a small but readily measurable chemisorption of ethane on nickel under these conditions. A rough estimate sug-

147

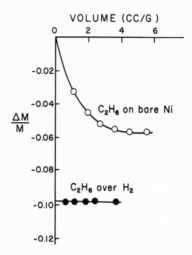

FIG. 60. Magnetization-volume isotherms for ethane and for ethane over preadsorbed hydrogen on nickel-kieselguhr, all at room temperature.

gests a minimum of eight bonds* for the average ethane molecule —thus abundantly establishing the dissociative mechanism. Precisely how the molecule is held to the metal is, as is well known from the work of Anderson and Kemball,[6] very much dependent on the experimental conditions.

If the metal surface is partly covered with pre-adsorbed hydrogen, then the fraction of chemisorbed ethane becomes negligible, although appreciable physical adsorption is still present—much of it no doubt on the silica support. This result is also in complete agreement with the infrared results of Eischens and Pliskin.[4]

2. The Associative Adsorption of Ethylene

In Eley's admirable review[2] of the catalytic hydrogenation of ethylene, he describes many different suggested modes of adsorption. Examples of these include the associative bond:

$$\begin{array}{cc} H_2C & \!\!\!\!-\!\!\!\!-CH_2 \\ | & | \\ Ni & Ni \end{array}$$

and several alternative dissociative mechanisms such as the following:

* Erroneously given as four by Selwood.[5]

HC=CH₂   H   H   HC=CH   H
  | | and | | | |
 Ni   Ni   Ni   Ni Ni   Ni

Other possibilities will suggest themselves.

Most of the evidence upon which these models were based consisted of kinetic and thermodynamic studies and calculations. These, together with deuterium exchange studies, such as those described by Kemball,[7] have not yet succeeded in establishing any one mechanism of adsorption—much less any one mechanism of hydrogenation. The difficulty is that there is too much evidence. Most authorities in the field would agree that the presence of acetylenic residues had been proved, all would agree that self-hydrogenation involving adsorbed ethylene can occur, all agree that adsorbed ethylene can poison nickel for the H_2—D_2 exchange reaction, and it has long been known that at higher temperatures carbiding may occur. All of these observations suggest a greater or less degree of dissociative adsorption, and yet the simple associative picture has an attractiveness which makes it hard to abandon. In the view of this writer we shall have to rely on physical methods if we are ever to gain a complete understanding of this problem.

The infrared absorption spectrum of ethylene adsorbed on nickel-silica has been studied by Pliskin and Eischens.[4,8,9] Unfortunately, the results, although valuable as an example of the method, do not lend themselves very well for comparison with other techniques. The reason for this is that the data thus far published have been obtained at, or slightly above, room temperature only, and that the pressure of adsorbate over the sample is, of necessity, always rather low. In spite of these limitations the infrared method has yielded an astonishing amount of information. On a nickel-silica sample evacuated at elevated temperature and hence presumably bare, the intensity of the absorption bands in the 3.4–3.5 micron region, and characteristic of C—H stretching vibrations associated with saturated carbon, is small compared with those bands observed when preadsorbed hydrogen is present. These results were obtained at 35°C, and they certainly suggest that most of the ethylene is adsorbed in a manner other than associative. On a surface partially covered with preadsorbed hydrogen there is also observed a band at 6.91 microns thought to be due to a scissorlike vibration of H—C—H, and hence showing the presence of at least two hydrogen atoms on the carbon. The situation on bare nickel is further complicated by the observation that the carbon atoms appear to be saturated, even though the ratio of hydrogen to carbon atoms is quite low. This seems to suggest either polymerization, or possibly bonding to two nickel atoms by each carbon as follows:

$$
\begin{array}{cccccc}
\text{H} & & \text{HC} & \text{---} & \text{CH} & \text{H} \\
| & \diagup & & \diagdown & \diagup \diagdown & | \\
\text{Ni} & \text{Ni} & \text{Ni} & \text{Ni} & \text{Ni} & \text{Ni}
\end{array}
$$

rather than

$$
\begin{array}{cccc}
\text{H} & \text{HC} & = & \text{CH} & \text{H} \\
| & | & & | & | \\
\text{Ni} & \text{Ni} & & \text{Ni} & \text{Ni}
\end{array}
$$

As mentioned above, Eischens and Pliskin found evidence for associative chemisorption of ethylene on a nickel surface which had been evacuated at room temperature after reduction, and hence probably was at least half covered with hydrogen.

Infrared studies of adsorbed ethylene have also been made by Pickering and Eckstrom[10] who found evidence for extensive, if not complete, dissociative adsorption on rhodium, and by Little et al.[11] who interpret their results as favoring the dissociative mode on palladium. Unfortunately, the transmission was poor for ethylene on nickel although some evidence was obtained for a ratio of $CH_2/CH_3 = 3$ on both nickel and palladium.

With the above brief introduction to a confused topic, we turn to the available magnetic data. A comparison has been made[12] of the change of magnetization produced in a reduced nickel-kieselguhr by hydrogen, on the one hand, and by ethylene, on the other. The temperature of adsorption and of measurement was 0°C, and the pressure was kept below 1 mm to diminish the possibility of self-hydrogenation, although this precaution proved to be unnecessary at this temperature. It was found that 8.21 cc (STP) of H_2 per gram Ni caused the magnetization to fall 11.2%, or a change of 1.36% per cubic centimeter of hydrogen.

At 0°C there is an appreciable van der Waals adsorption of ethylene. The procedure[5] was, therefore, as follows: first, the sample, was re-evacuated at 360°C, and ethylene to nearly 1 mm pressure was admitted at 0°C. The sample was then evacuated at 0°C through a Toepler pump, with no further change of magnetization and with pure ethylene as the desorbate. The volume of residual ethylene on the sample was then 1.68 cc (STP) C_2H_4 per gram Ni and this lowered the magnetization by 2.3%. This is a change of 1.34% per cubic centimeter of ethylene.

The magnetization change produced on nickel at 0°C by a molecule of ethylene is, therefore, the same as that produced by a molecule of hydrogen. This result is consistent with the view first advanced by Horiuti and Polanyi[13] that ethylene is adsorbed by the associative mechanism. The magnetic results alone do not exclude the mechanism CH_2=CH— plus H—, but it is well known that the maximum volume of ethylene which may be chemisorbed on nickel in the absence of self-hydrogenation[5,14,15] is only about one-third that of hydrogen. Any associative mechanism requiring two adjacent sites must obviously more quickly exhaust all possible sites than will a dissociative mechanism such as that indicated above. In this connection the remote possibility of carbon-carbon bond rupture (at this temperature) is excluded by the presence of ethane, but no methane, in the vapor removed by flowing hydrogen over the adsorbate. Our conclusion is, therefore, that under the conditions specified the principal mode of adsorption of ethylene is the associative. Further comment on this conclusion will be deferred until the more complicated results obtained under more drastic conditions have been presented.

3. Self-Hydrogenation, Dissociative Adsorption, and Carbiding

As the temperature of adsorption increases, the mechanism of adsorption for ethylene on nickel becomes increasingly complicated. An example of this is shown in Fig. 61 in which magnetization-volume isotherms are compared at 33.5° and at 100°C. At 33.5° the initial slope of the isotherm is about 1.8 times that of hydrogen while at 100° the slope is about 4.2 times that of hydrogen. These numbers, corresponding, respectively, to 3.6 and 8.4 bonds formed per molecule of ethylene adsorbed, can only mean that extensive dissociation takes place and that it rapidly becomes more severe at moderately elevated temperatures. There is probably also some effect of intrinsic activity because Broeder et al.[16] indicate an even greater change of magnetization when ethylene is adsorbed on an impregnated nickel-silica gel preparation.

In view of the large number of possible modes by which ethylene may undergo dissociative adsorption, it is probably idle to attempt

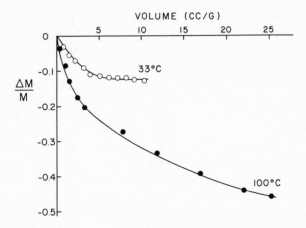

FIG. 61. Magnetization-volume isotherms for ethylene on nickel-kieselguhr at 33° and at 100°C.

any assignment of bond number to specific structure. In this respect the data obtained at an adsorption temperature of 28°C are instructive. The number of bonds apparently formed per molecule at this temperature is about three. This is probably an average caused partly by dissociation, including some carbon-carbon bond rupture, and possibly even including some polymerization of the dissociated fragments. Evidence that this is correct is obtained by passing hydrogen over the preadsorbed ethylene in an attempt to freeze out and identify the hydrogenation products. In a typical experiment it proved possible to remove 54% of the carbon, of which one-fifth was in the form of methane (establishing C—C bond rupture) and four-fifths as ethane. It seems probable therefore that even at 28°C almost half of the ethylene suffered no reaction much more serious than associative adsorption.

It might be thought that the maximum number of bonds which could be formed by adsorbed ethylene would be ten, representing two carbon atoms, or carbide ions, each held by three bonds, plus four Ni—H bonds, thus:

$$
\begin{array}{cccccc}
\text{H} & \text{H} & \text{C} & & \text{C} & \text{H} & \text{H} \\
| & | & \diagup | \diagdown & & \diagup | \diagdown & | & | \\
\text{Ni} & \text{Ni} & \text{Ni Ni Ni} & & \text{Ni Ni Ni} & \text{Ni} & \text{Ni}
\end{array}
$$

The initial slope of the magnetization-volume isotherm is actually fairly close to this value at temperatures slightly above 100°C, but still other complications arise. These are self-hydrogenation and complete conversion of the nickel to nickel carbide. Both reactions have been known for a long time.

On a nickel film, evidence of ethane in the vapor phase above chemisorbed ethylene is rapidly obtained at room temperature.[15] This occurs, presumably, by reaction of vapor phase molecules of ethylene with dissociated hydrogen atoms from preadsorbed ethylene. The ethane is generally discovered in the vapor phase as soon as the pressure becomes appreciable. There are obviously two convenient methods for studying this reaction. In one method the ethylene is admitted in batches and the products present in the vapor phase, or capable of being desorbed, are evacuated and analyzed. In the other method, ethylene is allowed to flow over the catalyst and the effluent gas is analyzed. This second method uses ethylene to sweep the metal surface free of self-hydrogenation products insofar as these are not themselves strongly adsorbed.

Nickel-silica preparations behave with respect to self-hydrogenation a little more sluggishly than do nickel films.[12] This difference is possibly due to the slower desorption of ethane, some of which must be physically adsorbed on the silica. In any event, ethylene over nickel-silica at 28°C, up to a pressure of 300 mm, yields no evidence of appreciable self-hydrogenation. This is found to be true both by the evacuation method and by the method of flowing ethylene over the sample at 1 atm; but if the experiment is repeated at quite moderately elevated temperature it will be found that a substantial fraction of the vapor over the catalyst is ethane. This is also found to be true if ethylene is allowed to flow over the sample in the neighborhood of 70°C—the effluent ethylene will contain ethane.

The reaction of self-hydrogenation has a peculiar effect on the quantity of ethylene which may be adsorbed on nickel. It has often been pointed out that the volume of ethylene which may be adsorbed on nickel is only one-half to one-third the volume of hydrogen which may be adsorbed on the same catalyst sample. Attempts have been made[14] to interpret this anomaly in terms of the supposedly favorable lattice spacings found on some nickel crystal faces and not on others. Beeck et al. reported that desorption of the ethane produced by self-hydrogenation would permit the adsorption of a substantial additional volume of ethylene on nickel. At the time Beeck's work was done there was no clear-cut method for differentiating between physical and chemical adsorption, and there was some uncertainty concerning how much additional ethylene could be said to be chemisorbed under these circumstances. The process has, however, been completely confirmed.[12] It will be described in some detail.

It was stated above that there is no evidence of appreciable self-hydrogenation on nickel-silica at room temperature. But if a catalyst sample is covered with 7.3 cc C_2H_4 per gram Ni at 33.5°C and then heated to 100°, it becomes possible to evacuate a considerable volume of ethane. If the sample is returned to 33.5° after the evacuation, it will be found that the nickel is now capable of sorbing a large additional volume of ethylene. Proof that this is chemisorption is obtained from the substantial additional loss of magnetization. We see, therefore, that the nickel is capable of taking up a volume of ethylene which even exceeds the maximum volume of hydrogen which may be adsorbed on the same sample. The requirement is that the temperature should be high enough so that a reaction may occur between ethylene and dissociated hydrogen and that the ethane should be removed.

As the adsorption temperature is raised still further, there appears abundant evidence that dissociation and other changes take place in the ethylene. At 87°C flowing ethylene causes the magnetization to fall to 58% of the initial value, and if hydrogen is then allowed to flow over the sample, it is found that the effluent gas contains a large amount of methane plus some ethane and a little higher hydrocarbon mixture. Comparable results may be obtained by evacuation and readmission of ethylene in the neighborhood of 100°C.

The final stage of dissociative adsorption is, of course, splitting off of all the hydrogen and rupture of the carbon-carbon bond. That this process occurs is shown by the following experiment. Ethylene was allowed to flow over the reduced nickel-silica sample at 126°C at a space velocity of about 1 per second. In 3 minutes the magnetization fell almost to zero, indicating almost complete conversion of the nickel to a nonmagnetic substance. If now hydrogen were permitted to flow over the sample, the hydrocarbon in the effluent was found to be almost pure methane, and the magnetization was in large part recovered. Evidence that this reaction of nickel and ethylene at moderately elevated temperature yields nickel carbide, Ni_3C, is found in the fact that heating the treated sample in vacuum at 355°C causes a substantial rise of magnetization (as measured at room temperature). This is in agreement with the observations of Hofer et al.[17] concerning the thermal stability of nickel carbide.

We may now summarize our views concerning the adsorption of ethylene on nickel. The widely divergent views held by many investigators are seen to be correctly attributed by Eley[2] to the complexity of the system due to the various possibilities for decomposition and polymerization. The supposedly limited ability of nickel to adsorb ethylene is due in part to the inhibition of the self-hydrogenation reaction at lower temperatures. The poisoning reaction is due primarily to dissociation and to carbiding. Schissler et al.[18] have shown that the poisoning effect of ethylene for the H_2—D_2 exchange reaction on nickel is much diminished if the ethylene is adsorbed below room temperature.

At lower temperatures the adsorption of ethylene is primarily associative. But the ethylene molecule is peculiarly sensitive to temperature and, presumably, to intrinsic activity of the catalyst. As the adsorption temperature rises there is a sharp rise in the production of ethane by self-hydrogenation. This reaction could hardly take place except by reaction of ethylene molecules with preadsorbed hydrogen formed by dissociation of the first ethylene molecules to strike the surface. That such dissociation takes place is confirmed by the progressive rise in the slope of the magnetization-volume isotherm for ethylene with rising temperature of adsorption.

As the adsorption temperature becomes still higher, it is found that the products formed by sweeping the surface with hydrogen contain an increasing proportion of methane, thus proving carbon-carbon bond rupture. Finally, at temperatures only moderately over 100°C, it is found that ethylene is capable of converting not only the surface but the whole mass of nickel into nickel carbide. This nickel carbide may readily be decomposed by hydrogen or, at somewhat higher temperature, by heat alone. This recovery of the nickel cannot, however, be achieved without some structural change in the catalyst such as growth of nickel particle size. The carbiding reaction is probably a chief offender in the poisoning action of adsorbed ethylene.

It will be noted that all of the several actions of ethylene on nickel have been proposed previously and have been the subject of investigation. The production of methane and the carbiding reaction were observed by Sabatier[19] many years ago. If there has been a divergence of views, it is because of failure to recognize that most of the proposed mechanisms possess an element of truth. If there is one area of disagreement, it is between the infrared absorption spectra which demands extensive dissociation at room temperature, and the magnetic data which subordinates dissociative adsorption to a minor role until somewhat higher temperatures are reached. We attribute this difference chiefly to differences in "intrinsic activity" between different catalyst preparations.

This section will be concluded with some brief remarks concerning possible sources of differences in "intrinsic activity." A perfect crystal face can have little relation to the kind of face exhibited to a reactant molecule by the typical nickel-silica catalyst preparation. The nickel particles in such preparations are, as we have seen, scarcely 50 A in diameter; many of them contain only a few hundred atoms, a substantial fraction of which are on the surface. Under such circumstances the various identifiable crystal faces must be of extremely limited area and most of them can consist of no more than a dozen or fewer atoms. At the edges of any such face there are atoms which lack much of the normal coordination in a metal lattice. Thus surface atoms in a perfect face might lack three nearest neighbors, those at an edge might lack six, and so on (see Fig. 62). It is hard to believe that this varied and almost

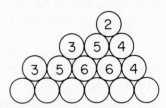

FIG. 62. Two-dimensional representation of the situation which may occur at the surface of an assembly of equivalent spheres. The number of nearest neighbors possessed by each sphere is shown.

random "degree of exposure" would not have a profound effect on the adsorptive ability of any such system and thus serve as a basis for differences in catalytic activity.*

The volume of the metal particles may also affect chemisorptive activity in the following way. When ethylene is adsorbed on nickel

* The idea expressed above is based on an article by Ehrlich and Turnbull[20] who show several excellent photographs of crystal models in which the number of nearest neighbors for various lattice surfaces is represented.

there may be observed[5] a fairly large thermal transient of the kind already described for hydrogen (p. 100). This heating of the nickel must vary inversely with the particle volume and, for a molecule as sensitive to adsorption temperature as ethylene, must certainly determine, at least in part, the degree to which dissociative adsorption takes place.

4. Remarks on the Hydrogenation of Ethylene

If the hydrogenation mechanism for ethylene on nickel is still obscure, we may, at least, see a reason for this state of affairs. If one grants that the ethylene must first be adsorbed (a circumstance which is by no means certain), then the numerous possible modes of adsorption make it difficult or impossible to formulate any one mechanism for all conditions. We have seen how the mode of adsorption depends upon temperature, ethylene pressure, intrinsic activity of the catalyst and, we may be sure, surface coverage with hydrogen or of ethylene. In principle almost any mode of adsorption not involving polymerization or C—C bond rupture might lend itself to the hydrogenation reaction. The various theories have been examined fairly recently by Eley[2] and by Horiuti.[21]

The idea put forward originally by Horiuti and Polanyi[13] was that the ethylene is associatively adsorbed and the hydrogen dissociatively adsorbed. Adsorbed C_2H_4 then reacts with adsorbed H to form adsorbed C_2H_5, which reacts with another adsorbed H to form C_2H_6 which is desorbed. A modification proposed by Twigg[22] calls for direct reaction of H_2 with adsorbed C_2H_4 to form adsorbed C_2H_5 and adsorbed H.

We shall not take space to discuss the many different mechanisms which have been suggested, but will mention one other. Jenkins and Rideal[23] suggested, in some contrast to the ideas mentioned above, that C_2H_4 from the gas phase reacts directly with adsorbed H to form adsorbed C_2H_5 or free C_2H_6. This idea is actually a modification of Beeck's theory[24] that molecular C_2H_4 reacts with adsorbed H on a surface already partly covered with acetylenic residues derived from preadsorbed C_2H_4. This last suggestion receives support from the kinetic studies of Wanninger and Smith.[25] Evidence for the presence of a half-hydrogenated state (adsorbed C_2H_5) is found in the infrared absorption studies of Eischens and Pliskin.[4]

The contribution of the magnetic method to this vexing problem

has thus far been small. It will be recalled that at room temperature on nickel-silica the ethylene molecule appears to be adsorbed
mostly by the associative mechanism, although there is also some
evidence for partial dissociation. In a typical run, the slope of the
magnetization-volume isotherm was found to be only a little greater
than that for hydrogen. Now if ethylene is preadsorbed at room
temperature and then hydrogen is added over the ethylene, we obtain the isotherm shown in Fig. 63.[5]

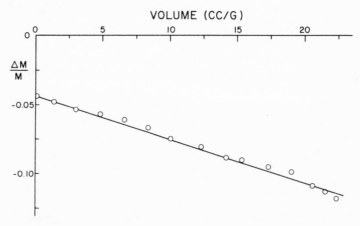

FIG. 63. Magnetization-volume isotherm for hydrogen over preadsorbed
ethylene on nickel-kieselguhr at room temperature.

There are two interesting features to this isotherm. The first is
that the isotherm is a straight line with a slope 2.4 times *smaller*
than that of hydrogen alone on the same surface.*
    The second point of interest is that the total volume of hydrogen
taken up is just enough to hydrogenate all the ethylene present
and to cover the surface with a monolayer of hydrogen. It is obvious that the reaction taking place is no mere replacement of a
Ni—C bond by a Ni—H bond. Such a reaction would yield no
change of magnetization.
    The most probable explanation for these results is that the
ethylene has covered only about one-third of the nickel surface
which is normally accessible to hydrogen. Then when the hydrogen

---

* This is a good example, in contrast to the results described on p. 111,
of hydrogen reacting with a preadsorbed molecule.

is added to the surface partially covered by preadsorbed ethylene, two molecules of hydrogen are adsorbed directly on to bare sites, one molecule of hydrogen is used to hydrogenate an ethylene molecule, and one molecule of hydrogen is adsorbed on the pair of sites vacated by the ethylene. This mechanism would lead to an isotherm slope for the hydrogen over preadsorbed ethylene just half that of hydrogen on a bare surface. But if we recall that some dissociation by the preadsorbed ethylene has occurred and that this gives an isotherm slope a little greater than that of hydrogen alone, then the expected ratio of slopes is quite near that actually found, namely, 1/2.4 instead of 1/2.

It will be noted that the conclusion reached above is not necessarily inconsistent with the transitory existence of a half-hydrogenated state, and it is not inconsistent with the view that ethylene is normally hydrogenated directly from the vapor state. All we have shown by the magnetic method is that over preadsorbed ethylene the hydrogen is used at random, partly for hydrogenation and partly for covering bare sites. Actual hydrogenation as, for instance, carried out routinely in organic chemistry generally calls for abundant hydrogen. This must mean that the metal surface is fully covered and that any unsaturated hydrocarbon must have great difficulty finding an appropriate adsorption site.* Viewed in this way the concept of direct reaction of ethylene from the vapor phase with adsorbed hydrogen atoms has considerable attractiveness. Particularly is this so when we have seen how readily, at quite moderate temperatures, it is possible to recover a fully carbided nickel catalyst. In view of all this we see that this kind of magnetic experiment does not tell us much concerning the practical aspects of catalytic hydrogenation. But it would appear that the magnetic monitoring of a nickel catalyst while it is actually functioning would be an instructive experiment, and there seems no reason why this type of study should not be extended to liquid phase reactions.

5. Acetylene

If attention has been lavished on the adsorption of ethylene this

* It is sometimes stated that because ethylene has a higher heat of adsorption on nickel it must be more strongly adsorbed. This is doubtless true on a surface otherwise bare, but it cannot necessarily be true on a surface already covered by hydrogen.

is not the case for acetylene. The subject is reviewed by Bond.[26] There can be no doubt that acetylene is chemisorbed on nickel in the room temperature region and higher. Conclusions based on reaction kinetics lead to the suggestion[27] that the adsorption mode is associative, thus:

$$HC\!\!=\!\!CH$$
$$\overset{|}{Ni} \ \overset{|}{Ni}$$

analogous to the associative mode for ethylene. While not denying this possibility under certain conditions, we must not be misled into thinking that this is the only possible mechanism on a bare nickel surface over a wide range of temperature.

There are some infrared absorption spectra data available for acetylene. Eischens[8,28] finds that the C—H stretching band characteristic of double-bonded carbon is absent in acetylene adsorbed on nickel. There is a band at 2940 cm^{-1} which is characteristic of saturated carbon. Other bands suggest two or even three hydrogens on each hydrogenated carbon atom, the spectrum being identical with that of adsorbed ethyl groups as found for ethylene adsorbed on a surface partly covered by hydrogen. This implies that the chemisorption of acetylene involves some degree of self-hydrogenation and this is supported by isotopic exchange experiments.[29] But some degree of polymerization is not excluded by these results. Infrared studies have also been made by Little et al.,[11] but here again transmission was poor on nickel. On palladium the evidence is that olefinic carbon is definitely present. This is interpreted as favoring the associative mechanism.

One sees from the results quoted above that we are not yet ready to estimate the number of bonds formed by acetylene under specified conditions, and much less ready to do so over a wide range of conditions. The only magnetic study on acetylene which has come to the writer's attention is that of Broeder et al.[16] It is scarcely possible to say what reliance should be placed on the data, which were obtained when the magnetic method was still in process of development. It appears, however, that a molecule of acetylene under the conditions of the experiment (these are not very clearly stated) is chemisorbed by 1.5 times as many bonds as a molecule of hydrogen. This is at least consistent with associative bonding plus a rather small amount of dissociation. Acetylene is clearly a molecule which deserves further investigation.

REFERENCES

1. B. M. W. Trapnell, in "Chemisorption" (W. E. Garner, ed.), p. 104. Academic Press, New York, 1957.
2. D. D. Eley, in "Catalysis" (P. H. Emmett, ed.), Vol. III, p. 64. Reinhold, New York, 1955.
3. J. R. Anderson, Revs. Pure and Appl. Chem. (Australia) 7, 165 (1957).
4. R. P. Eischens and W. A. Pliskin, Advances in Catalysis 10, 1 (1958).
5. P. W. Selwood, J. Am. Chem. Soc. 79, 3346 (1957).
6. J. R. Anderson and C. Kemball, Proc. Roy. Soc. A223, 361 (1954).
7. C. Kemball, Proc. Chem. Soc. p. 264 (1960).
8. W. A. Pliskin and R. P. Eischens, J. Chem. Phys. 24, 482 (1956).
9. R. P. Eischens, Z. Elektrochem. 60, 782 (1956).
10. H. L. Pickering and H. C. Eckstrom, J. Phys. Chem. 63, 512 (1959).
11. L. H. Little, N. Sheppard, and D. J. C. Yates, Proc. Roy. Soc. A259, 242 (1960).
12. P. W. Selwood, J. Am. Chem. Soc. 83, 2853 (1961).
13. J. Horiuti and M. Polanyi, Trans. Faraday Soc. 30, 1164 (1934).
14. G. H. Twigg and E. K. Rideal, Proc. Roy. Soc. A171, 55 (1939).
15. O. Beeck, A. E. Smith, and A. Wheeler, Proc. Roy. Soc. A177, 62 (1942).
16. J. J. Broeder, L. L. van Reijen, and A. R. Korswagen, J. chim. phys. 55, 37 (1957).
17. L. J. E. Hofer, E. M. Cohn, and W. C. Peebles, J. Phys. & Colloid Chem. 54, 1161 (1950).
18. D. O. Schissler, S. O. Thompson, and J. Turkevich, Advances in Catalysis 9, 6 (1957).
19. P. Sabatier, "Catalysis in Organic Chemistry" (translated by F. F. Reid), p. 153. Van Nostrand, New York, 1932.
20. G. Ehrlich and D. Turnbull, in "Metallurgical Society Conference" (N. Rhodin, ed.), p. 47. Interscience, New York, 1959.
21. J. Horiuti, J. Research Inst. Catalysis, Hokkaido Univ. 6, 250 (1958).
22. G. H. Twigg, Discussions Faraday Soc. 8, 159 (1950).
23. G. I. Jenkins and E. K. Rideal, J. Chem. Soc. p. 2490 (1955).
24. O. Beeck, Discussions Faraday Soc. 8, 118 (1950).
25. L. A. Wanninger and J. M. Smith, Chem. Weekblad 56, 273 (1960).
26. G. C. Bond, in "Catalysis" (P. H. Emmett, ed.), p. 109. Reinhold, New York, 1955.
27. J. Sheridan, J. Chem. Soc. p. 373 (1944).
28. R. P. Eischens, American Chemical Society Award in Petroleum Chemistry, Acceptance Address, April 1958.
29. J. E. Douglas and B. S. Rabinovitch, J. Am. Chem. Soc. 74, 2486 (1952).

CHAPTER X

Carbon Dioxide, Carbon Monoxide, Oxygen, and Nitrogen

1. Carbon Dioxide

Available information on the chemisorption of carbon dioxide on metals is meager. The data presented in this section will not be of much assistance in our understanding of the problem. They are given here as an aid in certain interpretations to be presented in the next section dealing with carbon monoxide.

Kokes and Emmett[1] have compared the adsorptions of reduced nickel oxide for carbon monoxide, carbon dioxide, and nitrogen. The criterion of chemisorption is that previously proposed by Emmett and Brunauer.[2] The total carbon dioxide sorption was determined to a fixed pressure at $-78°C$, the system was then evacuated at room temperature, and the sorption redetermined at $-78°$. Any difference between the volumes of adsorbate taken up in the first and second determinations at $-78°$ was taken as evidence for chemisorption of carbon dioxide. The results obtained in this study are not quite so explicit as might have been hoped. If interpreted literally they show that nickel is capable of chemisorbing carbon dioxide at $-78°$ to the extent of about 80% of the volume of nitrogen required to form a monolayer. But there is some question concerning the function of alkali impurity ions in the adsorbent. Furthermore, the criterion used is at best only an indication "once removed" of the electronic interaction which must occur with true chemisorption. We shall have reference to this matter again when we come to supposedly chemisorbed nitrogen.

The infrared absorption spectrum of carbon dioxide adsorbed on reduced nickel-silica at room temperature has been reported by Eischens and Pliskin.[3] At a pressure of 1.2 mm there is a strong

162

band at 6.4 microns and a weaker one at 7.1 microns. These are characteristic of the carboxylate ion, and suggest the following mode of adsorption:

$$
\begin{array}{cc}
O & O- \\
\diagdown & \diagup \\
 & C \\
 & | \\
 & Ni
\end{array}
$$

The only question we may have in this connection is whether the bands observed at this pressure may properly be attributed to carbon dioxide chemisorbed on nickel, rather than to physical adsorption. But Eischens and Pliskin also made studies on physically adsorbed carbon dioxide and their results seem to preclude this possibility on the nickel samples investigated. If Eischens and Pliskin have correctly interpreted the infrared data, it would appear that the band attributed by Fahrenfort et al.[4] to carbon dioxide refers to physical adsorption, as suggested, on the carrier.

Figure 64 shows data obtained by the author[5] for carbon dioxide

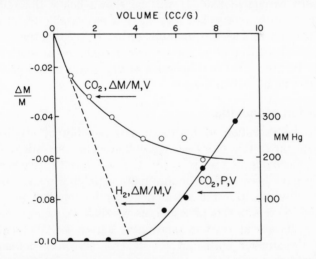

Fig. 64. Magnetization-volume and pressure-volume isotherms for carbon dioxide on nickel-kieselguhr at room temperature.

adsorbed on a reduced nickel-kieselguhr at 25°C. At a pressure of 237 mm the total sorption is nearly 9.0 cc (STP) per gram Ni, but evacuation at room temperature removes all but about 20% of this without change of magnetization. (An additional 10% may be removed by evacuation at higher temperature.) The total of definite chemisorption on the nickel is, therefore, quite small, although no smaller than that of ethane (p. 147) under the same conditions. The initial slope of the carbon dioxide isotherm is moderately greater than that of hydrogen.*

Our conclusion with respect to carbon dioxide is, therefore, that the molecule is held by at least two bonds and that for some reason which is obscure the maximum surface coverage is approximately one-eighth of that possible with hydrogen. In the case of ethane the demonstrated occurrence of dissociative adsorption, and the surface-covering ability of any molecule requiring two adjacent sites, makes it easy to understand why the apparent coverage is small. With carbon dioxide there is no obvious reason why the coverage should be less than half that of carbon monoxide. Actually it is barely one-sixteenth as great. We are forced to take recourse in the old idea that there are sites of greater and of less activity on nickel. Only the sites of greatest activity are able to chemisorb carbon dioxide. It will be shown below that adsorbed inert gases such as krypton may have a measurable effect on the magnetization of nickel (probably owing to polarization), and it might be thought that a similar effect could occur with carbon dioxide. But the effect of a molecule of krypton is only 16% of that of a molecule of carbon dioxide.

2. Carbon Monoxide

With the exception of hydrogen and possibly of ethylene, no adsorbate on metals has been more thoroughly investigated than carbon monoxide. Until recently the conclusions reached may scarcely have been said to be any more satisfying than those concerning ethylene. But the infrared absorption method has scored a notable success in this area and, as we shall see below, the magnetic results are at least in qualitative agreement. We shall first review the present status of the several pertinent nonmagnetic investigations.

Some important part of interest in the chemisorption of carbon monoxide is derived from a suggestion of Emmett and Brunauer[2] concerning the surface areas of supported metals, as of nickel on

---

* Dr. Umeda (personal communication) has recently confirmed the small chemisorption of carbon dioxide on nickel-silica, although the change of magnetization he observed is somewhat smaller than that reported here.

silica. While Emmett[1] still considers the method to be a useful one (as do many other investigators) yet Emmett himself has recognized that certain complications may arise, and some authors[6] do not hesitate to characterize the method as unreliable. It has been known for a long time[7-9] that the volume of carbon monoxide which may be sorbed on a nickel surface at room temperature is approximately twice the volume of hydrogen which may be chemisorbed under the same conditions. Physical adsorption is not important for carbon monoxide at room temperature, and hence these observations have led to the view that each surface nickel atom is attached in a "linear" mode, thus:

More recently Gundry and Tompkins,[10] on the basis of equilibrium and kinetic data, have shown that the H_2/CO ratio may be nearer $1/1.48$ than $1/2$, and from this they have concluded that two different modes of adsorption are possible. Qualitative support for this view has been obtained from electrical resistance studies on thin nickel films[11,12] and it has definitely been confirmed by the infrared studies which will be described in the following paragraphs.

Eischens et al.[13] showed that carbon monoxide on supported palladium, platinum, and nickel exhibits two major infrared bands at ~ 4.8 to 4.9 microns (~ 2180 to 2040 cm^{-1}), and at ~ 5.2 microns (~ 1930 cm^{-1}). The longer wavelength was assumed to correspond to a "bridged" CO molecule:

while the shorter was thought to correspond to a "linear" CO molecule attached to only one metal atom. The results obtained by Eischens et al. on nickel are not quite so nearly unambiguous as are those on the other metals, but there seems little escape from their conclusion that two modes of adsorption are actually possible. They find the band corresponding to the linear mode to be more pronounced at high surface coverage, and that corresponding to the bridged mode relatively more prominent at low coverage.

A further study of this problem has quite recently been published by (J. T.) Yates and Garland.[14] Most of their observations were made on reduced nickel supported on high-area alumina over a considerable range of nickel concentrations, but some observations were also made on silica-supported nickel. Improved design of the absorption cell has made it possible to work at pressures up to several millimeters of adsorbate, and the resolution is such that at least five bands are distinguishable.

The average nickel crystallite diameter was determined for those samples containing the largest (25%) concentration of nickel and estimated at 38 A by X-ray line width broadening.* Samples containing less nickel gave only diffuse X-ray diffraction—such samples were therefore classified as "crystalline," "semicrystalline," and "dispersed," but we cannot agree that anything approaching a true atomic dispersion was ever obtained, and it is not clear that the authors so intended. Yates and Garland then assign the several bands observed to possible structures as shown in Table XI.

TABLE XI

INFRARED ABSORPTION BANDS AND CONFIGURATIONS FOR CARBON MONOXIDE
ADSORBED ON NICKEL (AFTER YATES AND GARLAND)

Band	Wave number (cm^{-1})	Structure	Kind of Ni	Type of adsorption
A	1915	$\begin{array}{c} O \\ \| \\ C \\ / \quad \backslash \\ Ni \quad Ni \end{array}$	Cryst.	Very strong
C	2035	Ni=C=O	Cryst.	Very strong
B	1963	$\begin{array}{c} O \quad O \quad O \\ \backslash \quad \| \quad / \\ C \quad C \quad C \\ \backslash \quad / \backslash \quad / \\ Ni \quad Ni \end{array}$	Cryst.	Mod. strong
D	2057	Ni=C=O	Semicryst.	Mod. strong
E	2082	Ni=C=O	Disp.	Weak

* Our own experience is that this diameter may be rather low in terms of the magnetic method described on p. 44.

Interpreting the assignments given in Table XI literally, the authors then show that the adsorption of carbon monoxide on crystalline nickel sites occurs at very low pressure; adsorption on semicrystalline and dispersed sites occurs at higher pressure and this carbon monoxide is weakly bonded.

Assignment of band B to the particular bridged structure indicated in Table XI is given by Yates and Garland as a "possible" mode, but it is the most interesting in relation to our present problem. Band B forms on samples containing the higher concentrations of nickel at carbon monoxide pressures of several millimeters, and *after* bands A and C have been completely developed. The conclusions (of most direct concern to us) are, therefore, that the mode of adsorption is markedly dependent on the nickel particle size (and hence on the preparative procedure), that the mode is dependent on pressure (and hence to a degree on surface coverage), and that there is some rather convincing evidence concerning a bridged mode (B) forming between nickel atoms which already are involved with molecules adsorbed by the linear mode.

Another recent study by O'Neill and (D. J. C.) Yates[15] has extended studies of this type to high-area alumina, titania, and silica. The resolution obtained was not comparable with that reported by (J. T.) Yates and Garland, and no very significant change in the ratio of linear to bridged modes could be observed with changing surface coverage. Major differences were observed when the support was changed and these changes are tentatively ascribed to differences in the "electrical" properties of the supports in the manner discussed on p. 78. But in view of the demonstrated effect of particle size, and the lack of definite information concerning the nickel particle diameters, we must continue to reserve judgment on this matter of support influence.

The only available magnetic data on the problem of carbon monoxide adsorption are those obtained in the writer's laboratory.[16] A magnetization-volume isotherm for carbon monoxide adsorbed on nickel-kieselguhr at $-78°C$ is shown in Fig. 65. The isotherm for hydrogen under the same conditions on the same sample is also shown. The maximum pressure of carbon monoxide for the isotherm shown was 12 mm. If the pressure of carbon monoxide is increased beyond that shown in Fig. 65, a large additional sorption occurs, as expected from earlier work, but the magnetization isotherm suffers a rather abrupt change of slope, becoming much more

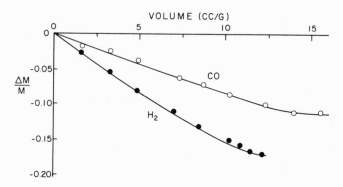

FIG. 65. Magnetization-volume isotherms for hydrogen and for carbon monoxide on nickel-silica at room temperature (after Den Besten).

nearly parallel to the volume axis. It is well known from the work of Yates and Garland, and others, that nickel tetracarbonyl is formed under these conditions. But the rate of formation at room temperature is so slow as not to interfere seriously with our interpretation of the data. Figure 66 shows isotherms obtained for

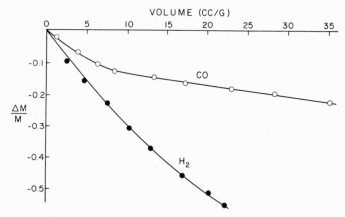

FIG. 66. Extension of the data shown in Fig. 65 to higher surface coverage of carbon monoxide (after Den Besten).

hydrogen and for carbon monoxide on a coprecipitated nickel-silica at 25°C. Pressure-volume isotherms are also shown. The nickel particle diameters, determined magnetically as described on p. 44 were 42 A for the nickel-kieselguhr (Fig. 65), and 25 A for the coprecipitate (Fig. 66).

Up to the point of change of slope the carbon monoxide cannot be evacuated at room temperature; evacuation at elevated temperature yields a substantial fraction of carbon dioxide formed, presumably, by disproportionation. But at higher surface coverages it is possible to evacuate some of the carbon monoxide as such at room temperature, and substantially all of it if the temperature is raised to 150°C. These changes will perhaps be clearer if presented in the form of an example. A nickel-kieselguhr at 25°C took up a total of 29.4 cc CO per gram Ni, of which 16.2 cc was beyond the change of magnetization slope. Then, on evacuation 5.0 cc was desorbed at 25°C, and 9.5 cc more at 150°, this latter fraction being about 95% carbon monoxide and 5% carbon dioxide. By evacuation at 400° it proved possible to recover nearly all of the carbon but most of this was in the form of carbon dioxide. During these changes the magnetization, as measured at 25°C, rose to about 95% of its initial value.

The initial slope of the magnetization-volume isotherm for carbon monoxide as shown in Figs. 65 and 66 is almost exactly half that of hydrogen. This strongly supports the linear, Ni≡C≡O, mode of adsorption, and the difficulty of removing this phase makes it consistent with the Yates and Garland band C. But beyond the change of slope it becomes much easier to evacuate carbon monoxide (although we do not know if the molecules are the same in the sense of last-on-first-off); and the magnetization slope becomes quite small. This suggests that the carbon monoxide at higher surface coverages is combining primarily with nickel atoms which are already bonded in the linear mode. The reaction might be represented thus:

$$
CO + \underset{Ni}{\overset{O}{\underset{\|}{\overset{\|}{C}}}} \quad \underset{Ni}{\overset{O}{\underset{\|}{\overset{\|}{C}}}} \rightarrow \quad \underset{Ni}{\overset{O}{C}} \quad \underset{Ni}{\overset{O}{C}} \quad \overset{O}{C}
$$

the bridged form being that designated as band B in the Yates and Garland scheme. It is, of course, entirely possible that this reaction causes a change in bond strength in the linear species, or desorption may occur directly from the bridged species.

Yates and Garland reported a notable dependence of chemisorption mechanism on nickel particle size, and this view is strikingly confirmed by Fig. 67. This shows an isotherm on nickel-kieselguhr

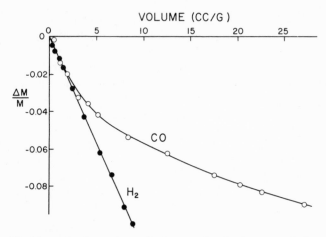

Fig. 67. Magnetization-volume isotherms for hydrogen and for carbon monoxide on nickel-kieselguhr which had been sintered and thus contained particles of nickel considerably larger than those used for Figs. 65 and 66. All at room temperature (after Fox).

which had been briefly sintered at 600°C to cause particle growth. The "magnetic" diameter of the nickel particles in this sample was 64 A. A diameter such as this brings the sample to the upper limit of applicability of the low-frequency AC permeameter. (The relaxation time becomes so large that the magnetization may no longer be directly proportional to the induced emf.) In spite of this drawback, it certainly appears from Fig. 67 that the initial slope of the isotherm is the same as that of hydrogen. If this is the case, then a bridged form at low coverage is clearly indicated.

Our conclusions with respect to carbon monoxide, based solely on the magnetic studies, are as follows:

(1) On small nickel particles the initial stage of chemisorption is probably mostly linear.

(2) At higher coverages it appears that two (or more) carbon monoxide molecules are adsorbed on one nickel atom in a fashion such as:

$$
\begin{array}{ccc}
O & O & O \qquad O \qquad O \\
\| & \| & \| \quad\ \| \quad\ \| \\
C & C & \text{or} \qquad C \quad C \quad C \\
\diagdown\ \diagup & & \diagdown\quad\diagup\ \diagdown \\
Ni & & Ni \qquad Ni
\end{array}
$$

(3) There is a strong dependence of the mode of adsorption on

nickel particle size. On larger particles the initial mode of adsorption appears to be bridged, that is to say, two nickel atoms are involved for every carbon monoxide molecule adsorbed.

(4) When two (or possibly more) carbon monoxide molecules are attached to one nickel atom, desorption of the monoxide as such becomes quite easy.

(5) Carbon dioxide is probably formed upon desorption from nickel atoms bonded to only one carbon monoxide molecule.

(6) The low chemisorption of carbon dioxide (p. 162) suggests that disproportionation of carbon monoxide does not occur until desorption is attempted.

The results and conclusions presented above are in gratifying qualitative agreement with those of Yates and Garland. The system is one which will lend itself to simultaneous infrared and magnetic measurements if, and when, such an experimental *tour de force* becomes possible.*

3. Oxygen

If the concept of chemisorbed oxygen has any physical meaning, it is clear that chemisorbed oxygen yields nothing to ethylene or carbon monoxide in complexity. It would appear that in addition to physically adsorbed O_2 molecules, we may have chemisorbed superoxide (O_2^-) ions, chemisorbed O^- ions (thus far nameless), and chemisorbed oxide (O^{2-}) ions.[17] Each of these may probably assume lattice positions so different as to constitute separate modes of adsorption. Our chief problems are then:

(1) Is there evidence for the chemisorption on nickel of oxygen which is in some manner physically different from a mere formation of nickel oxide extending some distance below the surface? And (2) if there is such a chemisorbed state (or states), what is its nature? That different forms of adsorbed oxygen may exist was shown by Roberts quite a few years ago.[18,19] (It will be noted that we carefully avoid the equally, or more, interesting problem of oxygen chemisorbed on nickel oxide).

We shall not learn the answers to these problems by examining the kinetic or equilibrium studies on the subject. Thus, in the

---

Note added in proof: J. W. Geus, A. P. P. Nobel, and P. Zwietering [*J. Catalysis* 1, 8 (1962)] have reported further magnetic studies on adsorbed carbon monoxide and other gases. Their results and conclusions appear to be in substantial agreement with those given above.

classical paper of Beeck et al.[8] it is correctly pointed out that the sorption of oxygen proceeds instantaneously to the extent of two molecules per lattice site, and it is correctly surmised that more than a surface reaction had taken place. As pointed out by Trapnell,[20] chemisorption of oxygen on metals like nickel and iron is rather more than likely to be followed by massive oxidation, although some evidence of a chemisorbed layer or at least of an oxide layer restricted to the actual surface, has been given by Zettlemoyer et al.[21]

The possibility that heats of adsorption would be of some help in differentiating between modes of oxygen chemisorption has been considered by Trapnell,[22] and by Klemperer and Stone.[23] The heat reported by the latter workers does nothing to alleviate the large (and almost unique) deviation found by Higuchi et al.[24] between the calculated and experimental heat of adsorption for the nickel-oxygen system.

The electrical conductivity of nickel films before and after exposure to low pressure of oxygen has been measured by Suhrmann and Schulz.[25] The conductivity decreases, at first abruptly and then almost imperceptibly, as more oxygen is taken up. The authors attribute this change to an extraction of electrons from the nickel by dissociated oxygen atoms. While not denying that such a process may take place, we can scarcely see any quantitative significance in these data. A study of potentially more interest is that of Lewis[26] on the K X-ray absorption edge of alumina-supported nickel and on the effect of adsorbed oxygen on the edge. The nickel crystallites were estimated by oxygen adsorption to be 30 A in diameter. The result of the study on the supposedly bare nickel is that the absorption edge is the same for these small particles as it is for massive nickel. This interesting result seems to raise doubt concerning the supposed effect of the support on the electronic structure of the supported metal (p. 77), and it also tends to confirm the view that essentially "clean" nickel surfaces may be prepared by the usual techniques of reducing and evacuating supported metal.

When Lewis adds oxygen to the alumina-supported nickel he finds that the absorption edge produced is remarkably similar to that of massive nickel oxide. This is interpreted to mean that the nickel-oxygen bond is ionic and that the nickel atom loses electrons to the oxygen. The interpretation suggested is that oxygen adsorption causes a diminution in the p character of the $4s$ band and an increase in that of the $4p$ band. We shall not find it easy to relate this interpretation to the magnetic data, for which we trust the reader is now reasonably well prepared.

The direct addition of oxygen to a reduced supported nickel sample almost, but not quite invariably, leads to a diminution of the magnetization.* A typical magnetization-volume isotherm for oxygen,[29] compared with hydrogen, on nickel-kieselguhr at 25°C is shown in Fig. 68. These results are in substantial agreement

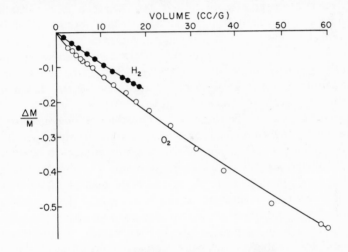

FIG. 68. Magnetization-volume isotherms for hydrogen and for oxygen on nickel-kieselguhr at room temperature (after Leak).

with the earlier results of Broeder *et al.*[28] obtained under similar conditions. Inasmuch as the conditions were not very different from those used by Lewis[26] and other investigators, we may probably safely conclude that this represents the condition postulated by Lewis, namely, a true formation of nickel oxide, but an oxide

---

* In the writer's first paper[27] on this subject it was stated that adsorbed oxygen causes an increase in the magnetization of nickel. It was thought that this represented an increase in the number of unpaired d electrons and that this was caused by electron transfer from metal to oxygen. Broeder *et al.*[28] were unable to repeat this observation. It will be shown below that oxygen may indeed cause an increase in the magnetization of nickel but that the author's interpretation in terms of d electron involvement is almost certainly wrong.

which has penetrated only a relatively short distance below the surface. It will be noted that the total volume of oxygen taken up by the nickel (Fig. 68) is quite large. This volume is to a degree dependent on the conditions of the experiment. The heat of sorption is large and, if the oxygen is admitted abruptly, the nickel becomes very hot and is more or less completely converted to bulk nickel oxide, although generally a small core of metal remains as shown by the failure of the magnetization to fall to zero.

It may be thought that as the slope of the oxygen isotherm is only moderately greater than that for hydrogen in the example given (Fig. 68), this indicates about two bonds formed, or at least two nickel atoms involved, per oxygen molecule adsorbed. This is perhaps what we might expect for purely ionic binding, but it is scarcely possible to be sure of this conclusion when we cannot even say for certain to what extent oxidation of the nickel may have penetrated below the surface.

Our next problem is then: Is it possible to demonstrate the formation of a chemisorbed oxygen layer which is somehow different from that described above? The answer appears to be "yes."

The difficulties encountered in studying chemisorbed oxygen must be due, in part at least, to the high heat of adsorption or reaction as the case may be. It may be thought that a different, perhaps transitory, mode of adsorption might be observed by keeping the rate of adsorption quite low as might, for instance, be achieved by holding the pressure, or partial pressure, of the oxygen at a low level. It may also be thought worthwhile to try lower temperatures of adsorption and to introduce the oxygen to the nickel surface indirectly in the form of nitrous oxide.[30] Nickel is an excellent catalyst for the decomposition of nitrous oxide and we may to some degree control the rate of decomposition by using a sintered, and hence less active, supported nickel adsorbent.*

If all the above devices are used, namely, lowered temperature,

---

* The adsorption of nitrous oxide on nickel is a rather instructive demonstration. On admission of a measured volume of the nitrous oxide to a reduced nickel catalyst it will be found that the pressure over the catalyst is such as to suggest that no adsorption has taken place. But analysis of the gas over the catalyst will show that it is no longer nitrous oxide but rather pure nitrogen. The catalytic decomposition of nitrous oxide by nickel is extremely rapid, the oxygen being chemisorbed and the nitrogen liberated. To the writer's knowledge no true chemisorption of molecular nitrous oxide per se on nickel has ever been demonstrated.

use of nitrous oxide as a source of oxygen, and sintered (and hence less active) nickel catalyst, then it is a simple matter to show that the magnetization of the nickel *increases* after the manner already described for hydrogen (p. 125) on nickel particles of diameter somewhat above the maximum exhibiting collective paramagnetism. Figure 69 shows a plot of magnetization versus gas volume ad-

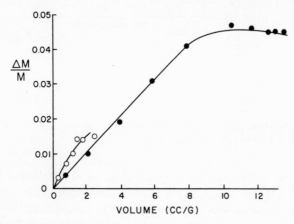

FIG. 69. Magnetization-volume isotherms for hydrogen (○) and for oxygen (●) on sintered nickel-kieselguhr at −78°C. The oxygen was derived from nitrous oxide (after Leak).

sorbed. This was obtained at −78°C in the manner indicated above.

The results shown in Fig. 69 do not in themselves prove anything other than that adsorbed oxygen may under certain circumstances produce an increase of magnetization similar to that produced by hydrogen and presumably for the same reason, namely, a decrease in the anisotropy of the particles. But if to an identical sintered catalyst sample one adds molecular oxygen in small increments at −78°, it will be found that the magnetization first shows a slight rise followed by a fairly substantial fall, as shown in Fig. 70. We may, therefore, conclude that there are indeed two modes of oxygen sorption. One of these we may designate as a true chemisorption (that is to say, a process restricted to the surface), and the other is probably a partial conversion to nickel oxide starting at the surface and penetrating a distance which depends on the experimental conditions. The first mode may be converted to the second merely by warming the sample.[30]

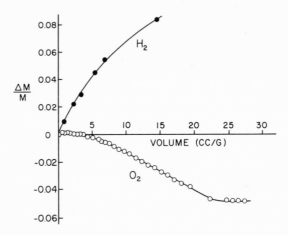

FIG. 70. Magnetization-volume isotherms for hydrogen and for oxygen on sintered nickel-kieselguhr at $-78°C$ (after Leak).

It will be noted that the positive slope of the isotherm for oxygen in Fig. 69 is about twice that of hydrogen. This suggests a bonding such as Ni—O—Ni, but it must be stated that the possible relation of these positive magnetization changes to saturation magnetic moments and bond types is quite speculative.

Some very convincing evidence concerning the existence of two or more modes of chemisorbed oxygen on nickel has been presented by Germer[31,32] and by Farnsworth,[33-35] both of whom used low-energy electron diffraction. These experiments were performed on crystal faces which were presumably as free from surface contamination as any yet produced, and they show that oxygen may form a crystalline surface lattice which may be quite readily distinguished from the oxide formed at higher "exposures" (i.e., longer *time* multiplied by *pressure*). It would be quite speculative to relate these different modes of adsorption to those which appear to be responsible for the magnetic behavior described above. The transition from one mode to another as revealed by electron diffraction occurs at "exposures" to oxygen of the order of 10^{-5} mm-sec. At first glance this behavior seems quite inconsistent with the observations of Leak and Selwood[29] in which the first chemisorbed species appears to be stable at "exposures" more conveniently expressed in millimeter-days rather than in millimeter-seconds. But this discrepancy may not be real. In a typical nickel-silica pellet the internal surface of the nickel is about 50 m². Any oxygen molecule entering the geometric boundary of the pellet must be instantly adsorbed and, as pointed out previously (p. 75), the partial pressure of a reactive gas *inside* the pellet may be many orders of magnitude less than the *external* pressure. But whether this is sufficient to relate the electron diffraction results to the magnetic cannot be said at the present time.

Our conclusions concerning adsorbed oxygen are that two or more modes of chemisorption have definitely been established. There is some evidence that the magnetic method is capable of distinguishing between at least two of these modes. There seems no doubt that d electrons in the nickel are directly affected by the adsorption of oxygen in either of these modes. It would, however, be helpful to have K edge absorption data for a system in which the oxygen was admitted in the form of nitrous oxide to a sintered sample at $-78°C$. This experiment would not be difficult and it might go a considerable distance in bridging the gap between the magnetic data and the low-energy electron diffraction results. It would also be helpful, although probably quite difficult, to measure heats of adsorption, or at least heats of interconversion, for the several modes of adsorption.

We may, perhaps, be permitted to remark at this point that the mechanisms of adsorption for even the simplest molecules seem almost infinite in their variety. No veteran of surface chemistry would maintain that any one mechanism, least of all *his* mechanism, was the *only* mechanism. We shall, for the most part, have to be content with trying to establish the commonest mechanism under a given set of conditions, or perhaps the mechanism of particular significance in some catalytic reaction. Even this modest achievement would be reason for satisfaction.

4. Nitrogen, Argon, and Krypton

The chemisorption of nitrogen on iron is a phenomenon of prime importance in the synthesis of ammonia. It is to be hoped that in due course the magnetic method will help to throw some light on this obscure process. For the present we shall have to confine our attention to nitrogen on nickel.

Over the years there have been several claims that nitrogen may be chemisorbed on nickel. The two studies of direct concern to us are those of Schuit and De Boer[36] and of Kokes and Emmett,[1] the former having used nickel-silica preparations quite similar to those used in many of the magnetic studies already described. Schuit and De Boer concluded that nitrogen may be chemisorbed on nickel because, primarily, the volume adsorbed at $-78°C$ was found to depend on the amount of nickel present and not on the total surface as determined by nitrogen adsorption at $-196°$. This work carries the implication that there is no chemisorption of nitrogen on nickel at $-196°$.

Kokes and Emmett, on the other hand, reached the conclusion that nitrogen may be chemisorbed on nickel at $-196°$. (If this is true, it tends to invalidate the Schuit and De Boer thesis.) The method used by Kokes and Emmett was to compare the total sorption of nitrogen at $-196°$ before and after evacuation at $-78°$. The difference was attributed to chemisorbed nitrogen. There was, however, evidence that some of the nitrogen thought to be chemisorbed at $-196°$ was removed by evacuation at $-78°$ and this introduces a degree of uncertainty concerning the actual volume of nitrogen which may be said to be chemisorbed.

There seems little reason to doubt the experimental facts presented by either group of investigators. If the effect described by Schuit and De Boer does not represent chemisorption, then we have no ready explanation for it. As for the conclusions reached by Kokes and Emmett, it should be pointed out that the method actually provides evidence which can at best be described as circumstantial. This is the kind of evidence which would provide useful confirmation provided that some direct evidence concerning electronic interaction were at hand. In the absence of such direct evidence we must refer once more to Barrer's comment[37] concerning the erroneous conclusions which may be based on sorption energy considerations in certain adsorbents of which silica gel is one. But whether this has any applicability to the unsupported nickel used by Kokes and Emmett we are not in a position to say.

Nitrogen sorbed on nickel has also been shown to affect the electrical conductivity[38] and to cause the appearance of a surface dipole,[39] but the effects are no larger than those produced by xenon[40] and cannot be considered proof of chemisorption at the present state of our understanding of these effects.

The effect* of adsorbed nitrogen on the magnetization of silica-

---

* In a preliminary paper,[27] the writer reported that adsorbed nitrogen caused an increase in the magnetization of nickel. This was, however, cor-

supported nickel has been investigated by the writer.[41] Nitrogen does, indeed, cause some loss of magnetization. The effect, which appears to reach a maximum in the neighborhood of −50°C is quite small, amounting to only a few per cent of the effect observed with an equal volume of adsorbed hydrogen. A magnetization-volume isotherm for nitrogen on nickel-silica at −78° is shown in Fig. 71. The isotherm was found to be reversible—only

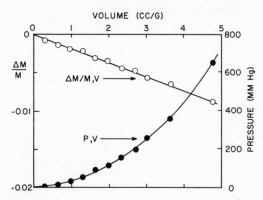

FIG. 71. Magnetization-volume and pressure-volume isotherms for nitrogen on nickel-kieselguhr at −78°C.

a trace of nitrogen remaining adsorbed after evacuation at −78° —and all the magnetization was recovered.

These results argue against any true chemisorption on the part of nitrogen under these conditions. It may be thought that perhaps the surface was already covered by preadsorbed nitrogen throughout the whole experiment. But the gas purification methods used make this contingency improbable, if not completely impossible. Even more convincing evidence is obtained by examining the effect of adsorbed inert gases. Helium is not adsorbed and shows no magnetic effect, even up to 140 atm.[42] Argon shows an effect comparable in magnitude with that of nitrogen, and krypton at −78° causes a loss of magnetization which is about 20% of that caused by hydrogen.

The reason for these magnetic effects of adsorbed inert gases (and probably also of nitrogen) may lie in the polarization[43] of

rectly shown by Broeder *et al.*[28] to have been due to failure to eliminate adsorbed hydrogen prior to admission of the nitrogen.

the adsorbate molecule which, in turn, causes some slight change of electron density in the several bands. This view is supported by the fact that krypton, the most polarizable molecule of the group, showed the largest effect. It would, however, be interesting to see if these inert adsorbates are capable of causing any change in the saturation magnetization as measured at liquid helium temperature and high field. There seems no obvious reason why this experiment should not be possible.

REFERENCES

1. R. J. Kokes and P. H. Emmett, *J. Am. Chem. Soc.* **82**, 1037 (1960).
2. P. H. Emmett and S. Brunauer, *J. Am. Chem. Soc.* **59**, 310, 1553 (1937).
3. R. P. Eischens and W. A. Pliskin, *Advances in Catalysis* **9**, 662 (1957).
4. J. Fahrenfort, L. L. van Reijen, and W. M. H. Sachtler, *in* "The Mechanism of Heterogeneous Catalysis" (J. H. De Boer, ed.), p. 23. Elsevier, Amsterdam, 1960.
5. I. Den Besten, P. G. Fox, and P. W. Selwood, *J. Phys. Chem.* **66**, 450 (1962).
6. J. H. De Boer, *in* "The Mechanism of Heterogeneous Catalysis" (J. H. De Boer, ed.), p. 5. Elsevier, Amsterdam, 1960.
7. I. Langmuir, *J. Am. Chem. Soc.* **38**, 2221 (1912).
8. O. Beeck, A. E. Smith, and A. Wheeler, *Proc. Roy. Soc.* **A177**, 62 (1940).
9. M. McD. Baker and E. K. Rideal, *Trans. Faraday Soc.* **51**, 1597 (1955).
10. P. M. Gundry and F. C. Tompkins, *Trans. Faraday Soc.* **52**, 1609 (1956); **53**, 218 (1957).
11. R. Suhrmann, G. Wedler, and H. Heyne, *Z. physik. Chem.* (*Frankfurt*) **22**, 336 (1959).
12. G. J. H. Dorgelo and W. M. H. Sachtler, *Naturwissenschaften* **46**, 576 (1959).
13. R. P. Eischens, S. A. Francis, and W. A. Pliskin, *J. Phys. Chem.* **60**, 194 (1956).
14. J. T. Yates and C. W. Garland, *J. Phys. Chem.* **65**, 617 (1961).
15. C. E. O'Neill and D. J. C. Yates, *J. Phys. Chem.* **65**, 901 (1961).
16. I. Den Besten, Doctoral Dissertation, Northwestern University, Evanston, Illinois, 1961.
17. F. S. Stone, *Advances in Catalysis* **13**, 1 (1962).
18. J. K. Roberts, *Proc. Roy. Soc.* **A152**, 445 (1935).
19. J. L. Morrison and J. K. Roberts, *Proc. Roy. Soc.* **A173**, 1 (1939).
20. B. M. W. Trapnell, "Chemisorption," p. 181. Academic Press, New York, 1955.
21. A. C. Zettlemoyer, Y. F. Yu, and J. J. Chessick, *J. Phys. Chem.* **64**, 12 (1960).
22. B. M. W. Trapnell, "Chemisorption," p. 100. Academic Press, New York, 1955.
23. D. F. Klemperer and F. S. Stone, *Proc. Roy. Soc.* **A243**, 375 (1957).
24. I. Higuchi, T. Ree, and H. Eyring, *J. Am. Chem. Soc.* **79**, 1330 (1957).

25. R. Suhrmann and K. Schulz, *Z. physik. Chem. (Frankfurt)* **1,** 69 (1954).
26. P. H. Lewis, *J. Phys. Chem.* **64,** 16 (1960).
27. P. W. Selwood, S. Adler, and T. R. Phillips, *J. Am. Chem. Soc.* **77,** 1462 (1955).
28. J. J. Broeder, L. L. van Reijen, and A. R. Korswagen, *J. chim. phys.* **54,** 37 (1957).
29. R. J. Leak and P. W. Selwood, *J. Phys. Chem.* **64,** 1114 (1960).
30. R. M. Dell, D. F. Klemperer, and F. S. Stone, *J. Phys. Chem.* **60,** 1586 (1956).
31. L. H. Germer and C. D. Hartman, *J. Appl. Phys.* **31,** 3085 (1960).
32. L. H. Germer, *Advances in Catalysis* **13,** 191 (1962).
33. R. E. Schlier and H. E. Farnsworth, *Advances in Catalysis* **9,** 434 (1957).
34. H. E. Farnsworth and J. Tuul, *J. Phys. and Chem. Solids* **9,** 48 (1959).
35. H. E. Farnsworth and H. H. Madden, *in* "Structure and Properties of Thin Films" (C. A. Neugebauer, J. B. Newkirk, and D. A. Vermilyea, eds.), p. 517. Wiley, New York, 1959.
36. G. C. A. Schuit and N. H. de Boer, *J. chim. phys.* **51,** 482 (1954).
37. R. M. Barrer, *in* "Chemisorption" (W. E. Garner, ed.), p. 91. Academic Press, New York, 1957.
38. R. Suhrmann and K. Schulz, *Z. Naturforsch.* **10a,** 517 (1955).
39. J. C. P. Mignolet, *Discussions Faraday Soc.* **8,** 105, 326 (1950).
40. R. Suhrmann, E. A. Dierk, B. Engelke, H. Hermann, and K. Schulz, *J. chim. phys.* **54,** 15 (1957).
41. P. W. Selwood, *J. Am. Chem. Soc.* **80,** 4198 (1958).
42. L. Vaska and P. W. Selwood, *J. Am. Chem. Soc.* **80,** 1331 (1958)
43. R. A. Pierotti and G. D. Halsey, Jr., *J. Phys. Chem.* **63,** 680 (1959).

Symbols Used More Than Once

H field strength

B magnetic induction

M magnetization

σ specific magnetization

κ susceptibility

χ susceptibility per gram

d density

T temperature, °K

Δ Weiss constant

μ magnetic moment

N number of particles of moment μ in a sample of unit volume

M_s saturation magnetization at temperature T

k Boltzmann constant

β Bohr magneton

χ_A susceptibility per gram-atom

μ_A atomic moment in Bohr magnetons

S spin quantum number

g splitting factor

J total (spin-orbital) quantum number

T_c Curie temperature

I_{sp} spontaneous magnetization

M_0 saturation magnetization at 0°K

$\bar{\mu}_A$ saturation moment per atom in Bohr magnetons

n number of atoms in a particle

v volume of a particle

V total volume of ferromagnetic particles in a sample

K anisotropy constant

M_r remanence

τ relaxation time

n_i number of particles of volume v_i in a sample

$\bar{v}$ average volume

α a constant

η demagnetization constant

V_T total volume (ferromagnetic plus nonferromagnetic) in a sample

M_{app} apparent magnetization

N_A number of metal atoms of average moment $\mu_A\beta$ in a sample of unit volume

ϵ change in $\bar{\mu}_A$ caused by one atom of adsorbed hydrogen per atom of adsorbent

N_H number of hydrogen atoms adsorbed on a sample of unit volume

E, E_0 permeameter emf readings

i permeameter current

n_H number of hydrogen atoms adsorbed on a particle

M' magnetization as altered by adsorbed atoms

μ_i moment of a particle of radius r_i

r_i radius of a particle

Δ_H number of hydrogen atoms adsorbed per unit area

θ fractional surface coverage

A_p surface area of a particle

$(n_H)_1$ number of hydrogen atoms adsorbed on a particle of volume v_1

$(N_H)_1$ total number of hydrogen atoms adsorbed on a sample containing particles of volume v_1

Author Index

Numbers in parentheses are reference numbers and indicate that an author's work is referred to although his name is not cited in the text. Numbers in italic show the page on which the complete reference is listed.

Subject Index

187